Contents

Chapter 1

Introduction

1.1 Vaporware

This article is about a computer industry term regarding product releases. For the company, see VaporWare (company). For the musical genre, see vaporwave.

In the computer industry, **vaporware** is a product, typi-

The U.S. Justice Department accused IBM of intentionally announcing its System/360 Model 91 computer (pictured) three years early to hurt sales of its competitor's computer.

cally computer hardware or software, that is announced to the general public but is never actually manufactured nor officially cancelled. Use of the word has broadened to include products such as automobiles.

Vaporware is often announced months or years before its purported release, with development details lacking. Developers have been accused of intentionally promoting vaporware to keep customers from switching to competing products that offer more features.[1] *Network World* magazine called vaporware an "epidemic" in 1989, and blamed the press for not investigating whether developers' claims were true. Seven major companies issued a report in 1990 saying they felt vaporware had hurt the industry's credibility. The United States accused several companies of announcing vaporware early in violation of antitrust laws, but few have been found guilty. *InfoWorld* magazine wrote that the word is overused, and places an unfair stigma on developers.

"Vaporware" was coined by a Microsoft engineer in 1982 to describe the company's Xenix operating system, and first appeared in print in a newsletter by entrepreneur Esther Dyson in 1983. It became popular among writers in the industry as a way to describe products they felt took too long to be released. *InfoWorld* magazine editor Stewart Alsop helped popularize it by lampooning Bill Gates with a *Golden Vaporware* award for the late release of his company's first version of Windows in 1985.

Vaporware first implied intentional fraud when it was applied to the Ovation office suite in 1983; the suite's demonstration was well received by the press, but the product was later revealed to have never existed.

1.1.1 Etymology

"Vaporware", sometimes synonymous with "vaportalk" in the 1980s,[2] has no single definition. It is generally used to describe a hardware or software product that has been announced, but that the developer has no intention of releasing any time soon, if ever.[3][4]

The first reported use of the word was in 1982 by an engineer at the computer software company Microsoft.[5] Ann Winblad, president of Open Systems Accounting Software, wanted to know if Microsoft planned to stop developing its Xenix operating system as some of Open System's products depended on it. She asked two Microsoft software engineers, John Ulett and Mark Ursino, who confirmed that development of Xenix had stopped. "One of them told me, 'Basically, it's vaporware'," she later said. Winblad compared the word to the idea of "selling smoke", implying Microsoft was selling a product it would soon not support.[2]

Winblad described the word to influential computer expert Esther Dyson,[2] who published it for the first time in her monthly newsletter *RELease 1.0*. In an article titled "Vaporware" in the November 1983 issue of *RELease 1.0*, Dyson defined the word as "good ideas incompletely implemented". She described three software products shown

Influential writer Esther Dyson (pictured here in 2008) popularized the term "vaporware" in her November 1983 issue of RELease 1.0.

at COMDEX in Las Vegas that year with bombastic advertisements. She stated that demonstrations of the "purported revolutions, breakthroughs and new generations" at the exhibition did not meet those claims.[3][6]

The practice existed before Winblad's account. In a January 1982 review of the new IBM Personal Computer, Byte *magazine* magazine favorably noted that IBM "refused to acknowledge the existence of any product that is not ready to be put on dealers' shelves tomorrow. Although this is frustrating at times, it is a refreshing change from some companies' practice of announcing a product even before its design is finished." [7] *Creative Computing* in March 1984 wrote of Coleco's delay in releasing the Adam that the company "did not invent the common practice of debuting products before they actually exist. In microcomputers, to do so otherwise would be to break with a veritable tradition." [8] After Dyson's article, however, the word became popular among writers in the personal computer software industry as a way to describe products they believed took too long to be released after their first announcement.[5] *InfoWorld* magazine editor Stewart Alsop helped popularize its use by lampooning Bill Gates, then CEO of Microsoft, with a *Golden Vaporware* award for the 18-month late release of Microsoft's first version of Windows in 1985. Alsop presented it to Gates at a celebration for the release while the song "The Impossible Dream" played in the background.[9][10]

"Vaporware" took another meaning when it was used to describe a product that did not exist. A new company named Ovation Technologies announced its office suite Ovation in 1983.[11] The company invested in an advertising campaign that promoted Ovation as a "great innovation", and showed a demonstration of the program at computer trade shows.[5][12] The demonstration was well received by writers in the press, was featured in a cover story for

an industry magazine, and reportedly created anticipation among potential customers.[12] Executives later revealed that Ovation never existed. The company created the fake demonstration in an unsuccessful attempt to raise money to finish their product,[11] and is "widely considered the mother of all vaporware," according to Laurie Flynn of *The New York Times*.[5]

Use of the term spread beyond the computer industry. *Newsweek* magazine's Allan Sloan described the manipulation of stocks by Yahoo! and Amazon.com as "financial vaporware" in 1997.[13] *Popular Science* magazine uses a scale ranging from "vaporware" to "bet on it" to describe release dates of new consumer electronics.[14] Car manufacturer General Motors' plans to develop and sell an electric car were called vaporware by an advocacy group in 2008.[15]

1.1.2   Causes and use

Late release

A product missing its announced release date, and the labeling of it as vaporware by the press, can be caused by its development taking longer than planned. Most software products are not released on time, according to researchers in 2001 who studied the causes and effects of vaporware;[10] the phenomenon is so common that Lotus' release of 1-2-3 on time in January 1983, three months after announcing it, amazed many.[2]

Software development is a complex process, and developers are often uncertain how long it will take to complete any given project.[10][16] Fixing errors in software, for example, can make up a significant portion of its development time,[10] and developers are motivated not to release software with errors because it could damage their reputation with customers.[10] Last-minute design changes are also common.[10] In 1986, the American National Standards Institute adopted SQL as the standard database manipulation language. Software company Ashton-Tate was ready to release their dBase IV database manipulation program, but pushed the release date back to add support for SQL. They felt their product would not be competitive without it.[12] As the word became more commonly used by writers in the mid-1980s, *InfoWorld* magazine editor James Fawcette wrote that its negative connotations were unfair to developers because of these types of circumstances.[17]

Vaporware also includes announced products that are never released because of financial problems, or because the industry changes during its development.[12] When 3D Realms first announced their video game *Duke Nukem Forever* in 1997, it was early in its development.[18] The company's previous game released in 1996, *Duke Nukem 3D*, was a critical and financial success, and customer anticipation for its sequel was high. As personal computer hardware speeds improved at a rapid pace in the late 1990s, it created an "arms race" between companies in the video game industry, according to *Wired News*. 3D Realms repeatedly moved the release date back over the next 12 years to add new, more advanced features. By the time 3D Realms went out of business in 2009 with the game still unreleased, *Duke Nukem Forever* had become synonymous with the word "vaporware" among industry writers.*[19]*[20] The game was revived and released in 2011. However, due to a 13-year period of anticipations and poor storyline, the game had primarily extremely negative reviews, except for PC Gamer, who gave it 80/100.

A company notorious for vaporware can improve its reputation. In the 1980s, video game maker Westwood Studios was known for shipping products late, but by 1993 it had so improved that, *Computer Gaming World* reported, "many publishers would assure [us] that a project was going to be completed on time *because* Westwood was doing it".*[21]

Early announcement

Announcing products early months or years before their release date,*[22] also called "preannouncing",*[23] has been an effective way by some developers to make their products successful. It can be seen as a legitimate part of their marketing strategy, but is generally not popular with industry press.*[24] The first company to release a product in a given market often gains an advantage. It can set the standard for similar future products, attract a large number of customers, and establish its brand before competitor's products are released.*[12] Public relations firm Coakley-Heagerty used an early announcement in 1984 to build interest among potential customers. Their client was a former employee of Atari who wanted to market his new arcade game, but his contract with Atari prohibited it until a later date. The firm created an advertising campaign including brochures and a shopping-mall appearance around a large ambiguous box covered in brown paper to "stall for time" until the game could be announced.*[2]

Early announcements send signals not only to customers and the media, but also to providers of support products, regulatory agencies, financial analysts, investors, and other parties.*[24] For example, an early announcement can relay information to vendors, letting them know to prepare marketing and shelf space. It can signal third-party developers to begin work on their own products, and it can be used to persuade a company's investors that they are actively developing new, profitable ideas.*[23] When IBM announced its Professional Workstation computer in 1986, they noted the lack of third-party programs written for it at the time, sig-

naling those developers to start preparing. Microsoft usually announces information about its operating systems early because third-party developers are dependent on that information to develop their own products.*[23]

A developer can strategically announce a product that is in the early stages of development, or before development begins, to gain competitive advantage over other developers.*[25] In addition to the "vaporware" label, this is also called "ambush marketing", and "fear, uncertainty and doubt" (FUD) by the press.*[23] If the announcing developer is a large company, this may be done to influence smaller companies to stop development of similar products. The smaller company might decide their product will not be able to compete, and that it is not worth the development costs.*[25] It can also be done in response to a competitor's already released product. The goal is to make potential customers believe a second, better product will be released soon. The customer might reconsider buying from the competitor, and wait.*[26] In 1994, as customer anticipation increased for Microsoft's new version of Windows (code-named "Chicago"), Apple announced a set of upgrades to its own System 7 operating system that were not due to be released until two years later. *The Wall Street Journal* wrote that Apple did this to "blunt Chicago's momentum".*[27]

A premature announcement can cause others to respond with their own. When VisiCorp announced Visi On in November 1982, it promised to ship the product by spring 1983. The news forced Quarterdeck Office Systems to announce in April 1983 that its DESQ would ship in November 1983. Microsoft responded by announcing Windows 1.0 in fall 1983, and Ovation Technologies followed by announcing Ovation in November. *InfoWorld* noted in May 1984 that of the four products only Visi On had shipped, albeit more than a year late and with only two supported applications.*[2]

Industry publications widely accused companies of using early announcements intentionally to gain competitive advantage over others. In his 1989 *Network World* article, Joe Mohen wrote the practice had become a "vaporware epidemic", and blamed the press for not investigating claims by developers. "If the pharmaceutical industry were this careless, I could announce a cure for cancer today – to a believing press." *[28] In 1985, Stewart Alsop began publishing his influential monthly *Vaporlist*, a list of companies he felt announced their products too early, hoping to dissuade them from the practice.*[5] *Wired* Magazine began publishing a similar list in 1997. Seven major software developers including Ashton-Tate, Hewlett-Packard and Sybase formed a council in 1990, and issued a report condemning the "vacuous product announcement dubbed vaporware and other misrepresentations of product availability" because they felt it had hurt the industry's credibility.*[29]

1.1.3   Antitrust allegations

Announcing a product which does not exist to gain a competitive advantage is illegal via Section 2 of the Sherman Antitrust Act of 1890, but few hardware or software developers have been found guilty of it. The section requires proof that the announcement is both provably false, and has actual or likely market impact.*[30] False or misleading announcements designed to influence stock prices are illegal under United States securities fraud laws.*[31] The complex and changing nature of the computer industry, marketing techniques, and lack of precedence for these laws applied to the industry can mean developers are not aware their actions are illegal. The U.S. Securities and Exchange Commission issued a statement in 1984 with the goal of reminding companies that securities fraud also applies to "statements that can reasonably be expected to reach investors and the trading markets" .*[32]

Several companies have been accused in court of using knowingly false announcements to gain market advantage. In 1969, The United States Justice Department accused IBM of doing this in the case *United States v. IBM*. After IBM's competitor Control Data Corporation (CDC) released a computer IBM announced the System/360 Model 91. The announcement resulted in a significant reduction in sales of CDC's product. The Justice Department accused IBM of doing this intentionally because the System/360 Model 91 was not released until three years later.*[33]*[34] IBM avoided preannouncing products during the antitrust case, but after the case ended it resumed the practice. The company likely announced its PCjr in November 1983 four months before general availability in March 1984 to hurt sales of rival home computers during the important Christmas sales season.*[35]*[36] The practice was not called "vaporware" at the time, but publications have since used the word to refer specifically to it. Similar cases have been filed against Kodak film company, AT&T, and Xerox.*[37]

US District Judge Stanley Sporkin was a vocal opponent of the practice during his review of the settlement resulting from *United States v. Microsoft* in 1994. "Vaporware is a practice that is deceitful on its face and everybody in the business community knows it," said Sporkin.*[38] One of the accusations made during the trial was that Microsoft has illegally used early announcements. The review began when three anonymous companies protested the settlement, claiming the government did not thoroughly investigate Microsoft's use of the practice. Specifically, they claimed Microsoft announced its Quick Basic 3 program to slow sales of its competitor Borland's recently released Turbo Basic program.*[37] The review was dismissed for lack of explicit proof.*[37]

1.1.4   See also

- List of vaporware
- Development hell
- List of commercial failures in video gaming
- Shovelware
- Concept car
- Osborne effect

1.1.5   Notes

[1] Vapour-ware definition of Vapour-ware in the Free Online Encyclopedia. Encyclopedia2.thefreedictionary.com.

[2] Shea (1984).

[3] Bayus; Jain; Rao (2001) p. 3.

[4] Prentice; Langmore (1994) p. 11.

[5] Flynn (1995), p. 1.

[6] Dyson (1983), pp. –6.

[7] Williams, Gregg (January 1982). "A Closer Look at the IBM Personal Computer" . *Byte*. p. 36. Retrieved 19 October 2013.

[8] Anderson, John J. (March 1984). "Coleco". *Creative Computing*. pp. 65–66. Retrieved 6 February 2015.

[9] Garud (1997); Ichbiah cited in Bayus; Jain; Rao (2001) p. 3.

[10] Bayus; Jain; Rao (2001), p. 5.

[11] Flynn (1995), p. 2.

[12] Jenkins (1998).

[13] Sloan (1997)

[14] "What's New". *Popular Science* (Bonnier Corporation): 15. 1 March 2007. ISSN 0161-7370. Retrieved 2010-04-15.

[15] Bersinger, Ken (5 April 2008). "Road for electric car makers full of potholes" . *Los Angeles Times*. Retrieved 2010-04-19.

[16] Johnston; Betts (1995).

[17] Fawcette (1985).

[18] Thompson (2009).

[19] Kesten, Lou (11 April 2009). "R.I.P. `Duke Nukem Forever'". *ABC News*. ABC News Internet Ventures. Archived from the original on 25 January 2012. Retrieved 2010-04-15.

[20] Kane, Yukari Iwatani (7 May 2009). "Duke Nukem Nuked" . *The Wall Street Journal Blogs* (Dow Jones & Company). Retrieved 2010-04-14.

[21] "Westwood Studios Partnership Hits Jackpot" . *Computer Gaming World*. 1993-08. p. 32. Retrieved 12 July 2014. Check date values in: |date= (help)

[22] Prentice; Langmore (1994) p. 2.

[23] Prentice (1996), p. 3.

[24] Prentice (1996), p. 4.

[25] Bayus; Jain; Rao (2001), p. 4.

[26] Haan (2003).

[27] Zachary; Carlton (1994)

[28] Mohen (1989).

[29] Messmer (1990).

[30] Bayus; Jain; Rao (2001), p. 11.

[31] Prentice; Langmore (1994) p. 15.

[32] SEC (1994) cited in Prentice; Langmore (1994) p. 17.

[33] Gerlach (2004).

[34] "IBM Antitrust Suit Records" . Hagley Museum and Library. Retrieved 2010-04-14.

[35] "I.B.M.'S Speedy Redirection" . *The New York Times*. 1983-11-02. Retrieved 2011-02-25.

[36] Freiberger, Paul (9–16 January 1984). "IBM indicates March as likely PCjr delivery date" . *InfoWorld*. p. 20. Retrieved 4 February 2015.

[37] Stern (1995).

[38] Yoder (1995) cited in Bayus; Jain; Rao (2001), p. 5.

1.1.6 References

- Flynn, Laurie (24 April 1995). "The Executive Computer" . *The New York Times* (The New York Times Company). ISSN 0362-4331. Retrieved 2010-04-14.

- Dyson, Esther (28 November 1983). "Vaporware" (PDF). *RELease 1.0* (Rosen Research): 5.

- Shea, Tom (7 May 1984). "Developers Unveil 'Vaporware'". *InfoWorld* (InfoWorld Media Group) **6** (19): 48. ISSN 0199-6649. Retrieved 2010-04-13.

- Bayus, Barry L.; Jain, Sanjay; Rao, Ambar G. (1 February 2001). "Truth or consequences: An analysis of vaporware and new product announcements" . *Journal of Marketing Research* (American Marketing Association) **38** (1): 3–13. doi:10.1509/jmkr.38.1.3.18834. ISSN 0022-2437.

- Johnston, Stuart J.; Betts, Mitch (13 February 1995). "Industry debates U.S. vaporware probe" . *Computerworld* (Computerworld): 2. ISSN 0010-4841.

- Jenkins, Avery (5 October 1998). "Long overdue; The reasons behind vaporware" . *Computerworld* (Computerworld): 10. ISSN 0010-4841.

- Gerlach, Heiko A. (2004). "Announcement, entry, and preemption when consumers have switching costs.(econometric analysis)". *RAND Journal of Economics* (The RAND Corporation) **35** (1): 184. doi:10.2307/1593736. ISSN 0741-6261.

- Haan, Marco A. (1 September 2003). "Vaporware as a Means of Entry Deterrence" . *The Journal of Industrial Economics* (John Wiley and Sons) **51** (3): 345–358. doi:10.1111/1467-6451.00204. ISSN 0022-1821.

- Stern, Richard H. (April 1995). "Microsoft and vaporware" . *IEEE Micro Magazine* (IEEE) **15** (2): 6–7. ISSN 0272-1732.

- Messmer, Ellen (22 October 1990). "Software firms form group to raise ethics". *Network World* (IDG Network World) **7** (43): 9. ISSN 0887-7661. Retrieved 2010-04-14.

- Mohen, Joseph (19 June 1989). "Seeking a cure for the vaporware epidemic" . *Network World* (IDG Network World) **6** (24): 32. ISSN 0887-7661. Retrieved 2010-04-13.

- Fawcette, James E. (10 June 1985). "Press' Vaporgate" . *InfoWorld* (InfoWorld Media Group) **7** (23): 5. ISSN 0199-6649. Retrieved 2010-04-14.

- Prentice, Robert (1996). "Vaporware: imaginary high-tech products and real antitrust liability in a post-Chicago world" . *Ohio State Law Journal* **57** (4). ISSN 0048-1572.

- Thompson, Clive (21 December 2009). "Learn to Let Go: How Success Killed Duke Nukem" . *Wired News* (Condé Nast Digital). Retrieved 2010-04-15.

- Sloan, Allan (28 April 1997). "Financial Vaporware" . *Newsweek* **129** (17): 57.

- Prentice, Robert A.; Langmore, John H. (1994). "Beware of varpoware: product hype and the securities fraud liability of high-tech companies" (PDF). *Harvard Journal of Law & Technology* (Harvard Law School) **8** (1). ISSN 0897-3393. Retrieved 2010-04-16.

- Zachary, G. Pascal; Carlton, Jim (7 March 1994). "Software rivals vying to define how PCs work". *The Wall Street Journal (eastern edition)* (Dow Jones & Company). ISSN 0099-9660.

1.1.7 External links

- Community Memory postings from 1996 on the term's origins crediting Ann Winblad and Stewart Alsop.

- *RELease 1.0* November 1983 a scanned copy of Esther Dyson's original article

Wired Magazine Vaporware Awards

See also: *Wired* (magazine)

1.2 List of vaporware

Vaporware is a product which is never released but never cancelled. The term "vaporware" can also refer to products that are released far behind schedule, or heavily-promoted products that do not actually exist. This list documents products which have been labelled as "vaporware".

1.2.1 Hardware

- Instabeat was a fitness app designed for swimming by Hind Hobeika. This waterproof tracker with a heart-rate monitor was designed to be mounted on swimming goggles so that it could give real-time feedback to swimmers on their performance levels. Instabeat has been accepting orders for well over 2 years and reportedly has a backlog of over 1,000 units. But no customers report having received their working units.

- Phantom was a console gaming system developed by Infinium Labs. A prototype was demonstrated in 2004, but its release was continually delayed and the company never announced that the product was cancelled. The company was accused of a pump and dump scam. It received the first place in "Vaporwares 2004" in Wired News.*[1]

- Lockitron was a device to allow a door deadbolt to be remotely controlled via Bluetooth or over the Internet. After a successful crowdfunding effort that raised over $1.5 million worth of pre-orders, Apigy has (as of February 2014) failed to deliver a product in substantial numbers.*[2]

1.2.2 Software

- Ovation was a highly promoted office suite. After demonstrations that were well received, it was later revealed that the product never existed. It is "widely considered the mother of all vaporware," according to Laurie Flynn of *The New York Times.*[3]

- Xenix is a discontinued version of the Unix operating system for various microcomputer platforms, licensed by Microsoft from AT&T Corporation in the late 1970s. The Santa Cruz Operation (SCO) later acquired exclusive rights to the software, and eventually superseded it with SCO UNIX (now known as SCO OpenServer). After the breakup of the Bell System AT&T started selling Unix. Microsoft, believing that it could not compete with Unix's developer, decided to abandon Xenix. The decision was not immediately transparent, and so Xenix gave birth to the term vaporware.*[4] An agreement was signed with IBM to develop OS/2, and the Xenix team (together with the best MS DOS developers) was assigned to that project. In 1987 Microsoft transferred ownership of Xenix to SCO in an agreement that left Microsoft owning 25% of SCO. When Microsoft eventually lost interest in OS/2 as well, it based its further high-end strategy on Windows NT.

Video games

- *Half-Life 2: Episode Three* - The trilogy of episodes following Half-Life 2 was intended to be concluded by the end of 2007. Although the first two episodes were released in a relatively timely fashion, the final installment never surfaced. Despite sporadic assurances from developer Valve Software that the sequel is in development, there has been no information about the game or when it may see a release. *Episode Three* has since become one of video gaming's most infamous cases of vaporware, and the time taken for any news to surface has also created rumours that the game, if released, will be a larger project - perhaps simply *Half-Life 3.*[5]

- *Commander Keen: The Universe Is Toast!* was supposed to be the third trilogy in the series, where the events were left off after the sixth episode, but it never actually had gotten out into the early stages of development, for id Software moved on to *Wolfenstein 3D* and then *Doom*. However, the author of the game often has commented that he would automatically make a new game if he ever gains back the intellectual property.*[6]

- *Descent* is a six degrees of freedom, first-person shooter game designed and developed by Parallax Software and published by Interplay Entertainment. The Wii port for the game was announced by Interplay and that it was intended to be released via the WiiWare service for the holiday season of 2010,*[7] but, however, no progress has been seen for the past 4 years, and it is now considered "abandoned", yet the company has not officially cancelled the port.

- *Tekken X Street Fighter* is a crossover 3D fighting game featuring characters from Bandai Namco Entertainment's *Tekken* series and Capcom's *Street Fighter* series, developed by Bandai Namco. The game was announced in July 2010 alongside Capcom's *Street Fighter X Tekken*, with some early character assets shown shortly thereafter.*[8]*[9] However, no screenshots or footage have been released since 2010, despite *Street Fighter X Tekken* and several *Tekken* projects having been released in the time since. As a result, many have speculated that the game was cancelled, though *Tekken* series producer Katsuhiro Harada has repeatedly maintained that the game is still in development.*[10]*[11]*[12]

1.2.3 Surfaced vaporware

Products which once were considered to be vaporware which eventually surfaced after a prolonged time:

- 3G*[13]

- Bluetooth*[14]

- *Duke Nukem Forever**[14] - Initial game development was announced in April 1997 with a scheduled launch of 1998, however the game experienced many delays and was not released until June 10, 2011, 15 years after initial development. See also Development of Duke Nukem Forever.

- *Daikatana**[15]

- Windows Vista (then, "Windows Code Name 'Longhorn'")*[15]

- Mac OS X, the long-awaited "next generation Mac OS" that finally shipped replacing the announced and later abandoned Copland, Gershwin and Taligent operating system attempts.*[16]

- *Warcraft III**[14]*[17]

- *S.T.A.L.K.E.R.: Shadow of Chernobyl* – Originally announced in 2001, the game experienced numerous delays.*[18] Beta builds of the final product have been

distributed to numerous game review sites.*[19] On 3 March 2007, THQ announced that the game had gone gold and was released on 20 March 2007, though it was leaked three days earlier.

- *Team Fortress 2**[15] was announced in 1999 and took 8 years to be released. With a complete change in gameplay and art direction, the North American release took place on 9 October 2007.

- *Black Mesa* was announced as a full remake of *Half-Life* in 2004. The first release date given by the developers was 2009, but development continued until 2012, when the first fourteen chapters were released on modding website Mod DB. It was greenlit for distribution on Steam on September 11 of the same year, before being released as an early access product on May 5, 2015, with numerous features from the mod improved upon in the Steam release. This remake was designed so as to become a better alternative for *Half-Life: Source* since it lacked new features other than the fact that it used the newer, revamped source engine. Due to its long development time (eleven years), the modification became notable for its delays, and dwindling updates on the status of its completion. The delays led to Wired awarding Black Mesa high spots on their "Vaporware Of The Year" lists in 2009 and 2010.*[20]*[21]

- *Doom* began development in May 2008,*[22] but was completely restarted in 2011 and later considered vaporware.*[23] It was finally showcased at E3 2015, and is scheduled for release in Q1/Q2 2016.*[24]

- *The Last Guardian* began development in 2007,*[25] and was formally announced at E3 2009.*[26] Very little information was released after this, before it was reintroduced at E3 2015; it is scheduled for release in 2016.*[27]

1.2.4 See also

- Vaporware

- List of failed and overbudget custom software projects

1.2.5 References

[1] Kahney, Leander (7 January 2005). "Vaporware Phantom Haunts Us All". Wired News. Archived from the original on 31 January 2006. Retrieved 2006-05-17.

[2] Lomas, Natasha (16 January 2014). "Lockitron Still Hasn't Shipped To Most Backers Over A Year After Its $2.2M Crowdfunding Effort". Tech Crunch. the keyless smart

lock that's designed to fit over your dumb deadbolt so you can lock and unlock your door with a smartphone remains so much sexy-looking vapourware

[3] Flynn, Laurie (24 April 1995). "The Executive Computer". *The New York Times* (The New York Times Company). ISSN 0362-4331. Retrieved 2010-04-14.

[4] Flynn, Laurie (24 April 1995). "The Executive Computer". *The New York Times* (The New York Times Company). ISSN 0362-4331. Retrieved 2010-04-14.

[5] "Vaporware 2010: The Great White Duke". Wired.com. Retrieved 30 July 2013.

[6] "A Look Back at Commander Keen". Retrieved July 18, 2015.

[7] Interplay Straps In with Descent for WiiWare

[8] Turi, Tim (2010-07-24). "Capcom Vs. Namco Is Street Fighter X Tekken". GameInformer. Retrieved 2010-08-05.

[9] "An Early First Look At Tekken X Street Fighter". 2010-08-19. Retrieved 2010-08-19.

[10] http://www.kotaku.com.au/2013/08/tekken-x-street-fighter-looking-for-the-right-time-to-release/

[11] http://www.polygon.com/2014/7/25/5937905/tekken-x-street-fighter-harada

[12] http://www.siliconera.com/2015/07/12/tekken-x-street-fighter-is-well-into-development-says-harada/

[13] Elisa Batista (March 6, 2002). "The Real Reason 3G is Vaporware". *Wired*. Retrieved June 24, 2014.

[14] "Vaporware 2000: Missing Inaction". Wired. 2001. Retrieved 2007-10-31. The bona fide beginning of the new millennium is almost upon us, but some things never change: The tech industry continues to whip up excitement by promising amazing new technologies, only to crush our spirits by delaying, postponing, pushing back or otherwise derailing the arrival of said goods – sometimes indefinitely.

[15] "Vaporware '99: The 'Winners'". Wired. 3 January 2000. Retrieved 2007-10-31. The last year of the last decade before 2000 has come and gone, but the Vaporware 1999 "winners" are still a dream to some, and a nightmare to others.

[16] Vaporware: Why Apple Doesn't Blog. Roughlydrafted.com (7 December 2006).

[17] "Vaporware 2001: Empty Promises". Wired. 7 January 2002. Retrieved 2007-10-31. Whatever you like to call it – the New Economy, the Dot-Com Economy, the Clinton Years – one thing is now clear about the period of prosperity that began in the mid-'90s and was snuffed out early last year.

[18] Top 10 Tuesday: Modern Vaporware. Pc.ign.com (11 April 2006).

[19] First impressions – S.T.A.L.K.E.R.: Shadow of Chernobyl. Eurogamer

[20] Calore, Michael (December 21, 2009). "Vaporware 2009: Inhale the Fail". *Wired (magazine)*. Condé Nast Publications. Retrieved September 22, 2012.

[21] Calore, Michael (January 3, 2011). "Vaporware 2010: The Great White Duke". *Wired (magazine)* (Condé Nast Publications). Retrieved September 15, 2012.

[22] Ocampo, Jason (May 7, 2008). "Doom 4 Announced". *IGN*. Ziff Davis. Retrieved July 5, 2015.

[23] Schreier, Jason (April 3, 2013). "Five Years And Nothing To Show: How Doom 4 Got Off Track". *Kotaku*. Gawker Media. Retrieved July 5, 2015.

[24] Pitcher, Jenna (June 14, 2015). "E3 2015: DOOM Release Date Announced". IGN. Retrieved June 14, 2015.

[25] Grifford, Kevin (2009-06-03). "Fumito Ueda Discusses Last Guardian". 1UP.com. Retrieved 2009-06-03.

[26] Clements, Ryan (2009-06-20). "E3 2009: Team ICO Presents The Last Guardian". IGN. Retrieved 2009-06-02.

[27] Sarkar, Samit; Crecente, Brian (2015-06-18). "The Last Guardian's incredible eight-year journey to the PlayStation 4". Polygon. Retrieved 2015-06-19.

Chapter 2

Infamous Examples

2.1 Amiga Walker

Not to be confused with the computer games Walker or Mind Walker.

The Amiga Walker (1996).

The **Amiga Walker**, sometimes incorrectly known as the *Mind Walker*, is a prototype of an Amiga computer developed and shown by Amiga Technologies in late 1995/early 1996. Walker was planned as a replacement for the A1200 with a faster CPU, better expansion capabilities and a built-in CD-ROM. The *Walker* was never released; Escom and Amiga Technologies went bankrupt, and only two prototypes were made.[*][1][*][2][*][3]

The case is unique and radically different from computers before it. The intention was also to make the motherboard available without the case so users could put it into a standard PC case. There were a number of other potential case designs of different sizes, the *Walker* motherboard could fit all of them; this allowed for expandability tailored to the user's requirements.[*][4]

Despite Amiga Technologies' bravery in producing an innovative case, it did not get a very good reception. Comparisons were made with vacuum cleaners, Darth Vader's helmet and even *Doctor Who's* electronic dog K-9.

2.1.1 Technical information

Specifications

The Amiga Walker motherboard.

- CPU:
  - Motorola 68030/33 MHz *(in the prototype version)*
  - Motorola 68030/40 MHz *(compared to 68020/14 MHz in A1200)*
- Chipset: AGA
- Memory:
  - 1 MB Kickstart ROM *(compared to 512 kB in the original Amiga 1200)*
  - 2 MB Chip RAM

- 4 MB Fast RAM (*only in the production version*)

- Drives:

  - internal CD-ROM

  - 1.44 MB internal floppy drive

- Realtime clock onboard

- Additional:

  - Amiga keyboard

2.1.2   See also

- Power A5000

- Amiga models and variants

2.1.3   References

[1] "The Amiga Walker". Nicholas Blachford. Retrieved 2008-11-26.

[2] "Amiga Technologies: Walker". Big Book of Amiga Hardware. Retrieved 2008-11-26.

[3] "Walker". Magazine AmiagOS et MorphOS. Retrieved 2008-11-26.

[4] "The Walker concept". Amiga history guide. Retrieved 2008-11-26.

2.1.4   External links

- Short video of an Amiga Walker prototype on YouTube

2.2   Asus Transformer Book Duet

The **Transformer Book Duet TD300**, was a 13.3 inch tablet computer that was developed by Asus.[1][2][3] The device used two operating systems interchangeably: Windows 8.1 by Microsoft, and Android 4.1 by Google.[2][3][4] The device featured a tablet screen and a detachable keyboard.[1][3][4] The device was reported to be cancelled due to opposition from both Google and Microsoft in mid-March, 2014.[5]

2.2.1   Design

The tablet portion of the Transformer Book Duet is 0.5 inches (1.3 cm) thick[6] with a 13.3 inch touch screen available in 1366×768 or 1920×1080 resolution.[1][7][8] The back of the screen is a rubberized black plastic, with the volume and power buttons embedded at the top right of the lid.[8] It has a detachable black chiclet keyboard with a key to switch between Windows and Android.[8]

The interchangeability of Android and Windows, coupled with the ability to switch between tablet and ultrabook form-factors have led some news sources to call the Transformer Book Duet a "four-in-one device".[9][10] Switching between the two operating systems takes about four seconds, according to Asus Chairman Jonney Shih.[4] The computer has an Intel Core i7 processor, and 4GB of RAM.[2][7] The tablet itself has a 128GB solid state drive, and the keyboard adds 1TB of hard drive storage.[2][11] The computer also features various ports including HDMI, LAN, one USB 3.0, and two USB 2.0.[11]

Daniel Griffiths of Forbes noted that this is not the first time Asus has experimented with "hybrid" devices. Asus has also developed the Eee PC, a laptop with a 7-inch display; and more recently, the PadFone, a smartphone marketed with companion tablet dock and keyboard dock accessories intended to improve functionality and battery life.[12]

2.2.2   Reception

Since its announcement at the Consumer Electronics Show on January 6, 2014,[1] the Transformer Book Duet has received a high degree of media attention from technology magazines and other mainstream news sources.

Vlad Savov of The Verge said that the Transformer Book Duet would work better as either a 10 or 11 inch device, rather than a 13-inch device, because Android "already struggles to fully capitalize on the real estate on screens of that size."[13] Joel Santo Domingo of PC Magazine said that the Transformer Book Duet is helpful for any consumer or business-person who needs to use both operating systems.[3]

Industry objection

In March 2014, *The Wall Street Journal* reported that because Microsoft and Google had both implemented policies which effectively ban the certification of devices which dual-boot both Android and Windows, the Transformer Book Duet would be cancelled, and Asus would pull its similar all-in-one desktops from the market. Both companies had reportedly objected to the concept of dual-OS devices of this nature as early as January 2014. Prior to CES, an analyst believed that Microsoft was discouraging manufacturers from releasing such devices because they would dilute Windows 8 and Windows Phone's software ecosystem (which Microsoft was reportedly planning to unify). He also speculated that Microsoft would penalize OEMs by forfeiting discounts on Windows licenses and refusing to provide them with financing for marketing. Additionally, even though Android is an open source operating system that is freely available, Google's application suite is proprietary, and can only be licensed for devices which are approved by the company. This would prevent such devices from including access to Google Play, Android's primary application store.*[14]*[15]*[16]*[17]

2.2.3 See also

- Samsung Ativ Q

2.2.4 References

[1] Swamy, Rohan (7 January 2014). "CES 2014: Asus Transformer Book Duet launched with dual-boot Android, Windows". New Delhi Television Limited. Retrieved 7 January 2014.

[2] Ingraham Nathan (6 January 2014). "The Asus Transformer Book Duet hybrid can instantly switch between Windows and Android". The Verge. Retrieved 6 January 2014.

[3] "Asus Transformer Book Duet Dual Boots With Windows 8.1, Android". PCmag. 6 January 2014. Retrieved 7 January 2014.

[4] Limer Eric (6 January 2014). "Asus Transformer Book Duet: One Laptop, Two Tablets, All The Crazy". Gizmodo. Retrieved 6 January 2014.

[5]

[6] Osborne Joe (6 January 2014). "Asus Transformer Book Duet is a hybrid laptop in more ways than one". techradar pro. Retrieved 11 January 2014.

[7] Mack, Eric (11 January 2014). "ASUS' Transformer Book Duet: Home to both Android and Windows 8.1". Gizmag. Retrieved 12 January 2014.

[8] Cunningham, Andrew. "Hands-on: Asus' Transformer Book Duet and the clunky "Dual OS" feature". ars technica. Retrieved 11 January 2014.

[9] Summerson, Cameron (7 January 2014). "Hands On With The ASUS Transformer Book Duet". Android Police. Retrieved 11 January 2014.

[10] Triggs, Robert (10 January 2014). "ASUS Transformer Book Duet 4-in-1 hands on preview: video". Android Authority. Retrieved 11 January 2014.

[11] "Asus Transformer Book Duet TD300 tablet/laptop hybrid runs Windows 8.1 and Android". CNET. 6 January 2014. Retrieved 7 January 2014.

[12] Griffiths Daniel (7 January 2014). "Asus Announces Transformer Book Duet - Windows And Android, Together At Last. Wait, What?". Forbes. Retrieved 7 January 2014.

[13] Savov Vlad (7 January 2014). "The Transformer Book Duet combines Windows with Android, tablet with laptop". The Verge. Retrieved 7 January 2014.

[14] "Microsoft and Google ruin Intel's plan for dual-OS tablets". The Verge. Retrieved 17 March 2014.

[15] "Google and Microsoft are out to stop dual-boot Windows/Android devices". Ars Technica. Retrieved 17 March 2014.

[16] "Intel plans a CES coup: Android and Windows in the same computer". The Verge. Retrieved 17 March 2014.

[17] "Google's iron grip on Android: Controlling open source by any means necessary". Ars Technica. Retrieved 2013-12-08.

2.3 Beyond Good & Evil 2

Beyond Good & Evil 2 is a video game in development by Ubisoft Montpellier and published by Ubisoft. It is the sequel to the 2003 video game *Beyond Good & Evil*. Intended as the first part of a trilogy, the original *Beyond Good & Evil* won critical acclaim but failed to gain commercial success. As such, the status of *BG&E2* was unknown for several years until it was unveiled at Ubidays 2008 in the Louvre in Paris, France, on May 28, 2008.

Since its first revelation *BG&E2* 's development has been characterized in the media by uncertainty, doubt and rumours about the game's future. On May 28, 2010, a French website claimed to have insider information of Michel Ancel leaving Ubisoft Montpellier, thus putting *BG&E2* on hold.*[2] However this was denied by Ubisoft representatives shortly thereafter.*[3] Appearing at the Montpellier in Game conference on June 25, 2010, Ancel stated that the game is in development and that they are experimenting with new development practices to keep the development team small and preserve its artistic spirit. As such, the development will take "a while", but Ancel encouraged listeners to be patient.*[4] In 2011, it was revealed that Ancel and his team took a break from the development of the game to work on *Rayman Origins*.*[5]

Speaking at an exclusive Ubisoft dinner in 2011, Ancel stated that *BG&E2* was far into development and targeting a release on the next generation of consoles.[*][6]

2.3.1 Development

The original *Beyond Good & Evil* game was meant to be the first of a trilogy of games. Michel Ancel has stated that he wrote the story of the *Beyond Good & Evil*-universe to be longer than what is included in the first game, but because of the poor sales of the original game, Ubisoft was reluctant to invest in a sequel.[*][7]

Michel Ancel first gave a hint about *Beyond Good & Evil 2* in an interview with *Nintendo Power*, where he confirmed that he was working on a new project that means a lot to him. He also talked about Jade (the protagonist from the original game) and said that he really hopes that she will continue to keep her values and her personality.[*][8]

Many rumours about a possible sequel appeared after the release of the first game, all of which seemed to be false until May 2008 when Ubisoft revealed a teaser trailer for Ubisoft Montpellier's new project which featured characters and remixed music from the original *Beyond Good & Evil*.[*][9]

Speaking to a French magazine, Ancel cited that *Beyond Good & Evil 2* is a very personal project to him, and that he and his team wish to dedicate as much time as they feel necessary to it, in order to achieve the result they desire. He also affirmed that the gameplay will reflect aspects more in tune with the third-person style of *Assassin's Creed*, than the first-person perspective seen in *Mirror's Edge*.[*][10]

Announcement

A screenshot from the teaser trailer, showing Pey'j in the foreground.

Michel Ancel first talked about *Beyond Good & Evil 2* being in production in an interview with the French magazine Jeux Vidéo Magazine, where he stated that the game

A screenshot from a leaked trailer showing Jade being pursued through an urban setting in a third-person perspective.

has been in pre-production for a year, but was yet to be approved by Ubisoft.[*][11] Less than two weeks later, a teaser trailer of the game, recorded entirely inside the game's engine, was shown at Ubidays. However, the game was presented only as the next project of Michel Ancel and Ubisoft Montpellier; no specific details, release date or title was announced. The trailer, showing off the highly advanced graphics of the game engine, depicts a broken-down hovercar on the shoulder of a desert road. Beside the vehicle sits Pey'j, one of the supporting characters from the original game. A woman resembling Jade, the protagonist, sits on the trunk in the background. Not much is revealed about the game itself.[*][12][*][13] The Jeux Vidéo Magazine article also stated that *Beyond Good & Evil 2* is under development for PlayStation 3 and Xbox 360, but this is not confirmed.[*][14]

On December 18, 2008, at the VGL event in Paris, Ancel stated that *Beyond Good & Evil 2* has been under development for a year and a half and that the development team have received total freedom from Ubisoft, giving them the opportunity to make the game how they want.[*][15]

In an interview on March 21, 2009, Ancel stated that the game was still in pre-production and that they would not start actual development until they had decided on all the tools and processes needed to realize their vision of the game.[*][16]

On May 8, 2009, a video was released onto the Internet showing a character resembling Jade running through a city, dodging its police force.[*][17] On July 22, 2009, Laurent Detoc, CEO of Ubisoft North America, confirmed that the video had been leaked but said that this was not done intentionally. He continued to state that they had started working on *Beyond Good & Evil 2* because they didn't want to abandon the IP as it has a cachet and authenticity about it, but also emphasized that although they are definitely working on the game it doesn't necessarily mean it's ever going to be released; that's something the future will decide.[*][18] The authenticity of the trailer was confirmed by Michel Ancel at E3 2011.[*][19]

During the Gamescom convention in August 2009, Ubisoft

representatives stated that the game's development was officially on-hold.[20] However, shortly after the convention Ubisoft stated in an interview that work on the title had been neither abandoned nor put on hold.[21] When fans asked for clarifications regarding the mixed signals coming from Ubisoft, a UK forum manager stated that it was not technically possible for the game to be delayed since it had not actually been officially announced.[22] In November 2009, Geoffroy Sardin, the General Director of Ubisoft France, said that the rumours about the game being put on hold were false and that development was still on-going.[23] This was confirmed again in January 2010 by Ubisoft representatives talking to IGN.[24]

On May 28, 2010, (exactly two years after the initial announcement) the French website Wootgaming claimed to have insider information of Michel Ancel leaving Ubisoft Montpellier due to internal problems with the development of *BG&E2*, and consequently putting the game's development on hold.[2] The rumours were later denied, however, by Ubisoft PR representatives, as well as sources within the studio, and reassured that *BG&E2* is still under development.[3]

On June 25, 2010, Michel Ancel appeared at the Montpellier in Game conference and spoke about *BG&E2*. He assured that the game in fact still is in development, but is being worked at by a relatively small number of people and thus will take a while to finish. Ancel wants to keep the development team as small as possible to preserve *BG&E2* 's artistic spirit and avoid it becoming just a commercial product. The team has been working on tools that will enable developers to create 3D-games in a similar way as the tools being used for the 2D-game *Rayman: Origins*, which has enabled a small sub-team of only five people to create the game's levels.[4]

On August 19, 2010, Yves Guillemot, CEO of Ubisoft, told Kotaku that work continues on *Beyond Good & Evil 2* and had never stopped. *We were, as we are, working on the game. What is very important with this next product is that it will be perfect.*[25]

The HD re-release of the original game was released in March 2011 on Xbox Live Arcade and in June 2011 on PlayStation Network. The producer of the updated version, Wang Xu, said to "keep the faith" for the sequel. He has also said that he does not know if the HD version is used to test how customers feel about the franchise.[26]

On June 7, 2011, Ancel revealed that he plans to develop *BG&E2* for the next generation of consoles, which were yet to be announced.[19]

In May 2012, it was announced by Michel Ancel that the game had entered an 'active' development stage specifically for next-gen consoles, with them working on the in-game

engine. They also previewed a screenshot of the in-game graphics.[27]

On July 31, 2014, Ubisoft officially confirmed that Beyond Good & Evil 2 is in active development at the moment.[28] The exact words from publisher's press person were:

> *In many ways, BG&E is an inimitable game - it appeals to all generations of gamers and is an inspiration behind many of Ubisoft Montpellier's past and future games. It's still far too early to give many details about this new title, but what we can say is that while Michel and the team at Ubisoft Montpellier are working with the core tenets of BG&E, they're developing something that aspires to push past the boundaries of a proverbial sequel and leverages next-gen technologies to deliver a truly surprising, innovative and exceptional game. The entire team is excited about the direction this extremely ambitious project is taking, and we'll have more to share later, as it progresses.*

This statement was made shortly after Ancel's announcement of starting a project of his own game studio dubbed Wild Sheep Studio which employs 13 people, including Michel himself.

In June 2015, it was announced that *Beyond Good & Evil 2* was being play-tested at Ubisoft Montpelier.[29]

Cast

It is unknown if the cast of the original *Beyond Good & Evil* is coming back to work on *Beyond Good & Evil 2*. However, it is confirmed that composer Christophe Héral, who scored the music for the original game, will be returning.[30] In an interview with Jeux Vidéo Magazine, Ancel stated that a small team had already been working on the pre-production for a year.[31]

Target audience

In an interview with Next-Gen, Ubisoft CEO Yves Guillemot stated that *Beyond Good & Evil 2* will be made more accessible to the new generation of players, in an effort to make sure the sequel does not suffer the same commercial failure the original did.[32] Guillemot clarified that statement later on, saying that they do not intend to make the game more casual game oriented.[33]

2.3.2 Plot and setting

In the interview with Jeux Vidéo Magazine, Michel Ancel stated:

"[*Beyond Good & Evil 2* will] be in continuity with the first game, with a big variety of levels, lots of emotion in the gameplay, and characters we care about. This time we are dealing with planet's future, and the relationship with animals..." [*][11]

On April 3, Eurogamer.net published a translated interview with Ancel talking about the inspirations behind *Beyond Good & Evil* and its sequel:

"It was a mix of experiences. It was a phantasm to create an adventure game, a universe too. It was the game I wanted to create for a long time. [...] There were a lot of inspirations: the Miyazaki universe, my own inspirations, politics and the media; the theme of September 11 – the CNN show with army messages and the fear climate. And it was a mix from other universes." [*][34]

He also answered the question of whether or not the game will be a direct sequel to the first game:

"Yes! But it is better to discover that when you play, ha ha! But it is clear we want to continue the story – we won't create characters, story. The story continues and we'll react on important events of the first." [*][34]

Ancel also talked about the sequel being more immersive and complex compared to the original, largely because of the possibilities of current-generation hardware, but as a result will take longer to develop.

"We'll conserve the spirit of the first game, but the form will change. For [BG&E], we wanted to create a more cinematographic game, but we didn't have the technical ability to do that. So we simplified. It was fun, because I remember we had to work for Sony to demonstrate how original our game was in order to get the development kit. We wrote BG&E like we were in a film. After, we understood it would be very tricky to reach that immersion on that console! But now, with the next-gen consoles, we feel it is possible to develop the game we thought of for the first. [...] It is very difficult to answer [when the game's going to be released]: there is an unknown aspect." [*][35]

2.3.3   References

[1] JC Fletcher (2009-02-19). "Ubisoft programmer lists 'Rabbids Go Home' on (raving) resume". Joystiq. Retrieved 2009-05-11.

[2] Nathan Grayson (2010-05-28). "Rumor: Michel Ancel leaves Ubisoft Montpellier, Beyond Good & Evil 2's future in doubt.". VG247. Retrieved 2010-05-29.

[3] "Michel Ancel quitterait Ubisoft? C'est faux!" (in French). Gameblog.fr. 2010-05-28. Retrieved 2010-05-29.

[4] Oli Welsh (2010-06-25). "Ancel using small team to make BG&E2". Eurogamer. Retrieved 2010-06-26.

[5] *Game Informer*, Issue 218, June 2011

[6] Matt Martin (2011-06-06). "No Beyond Good & Evil 2 this generation". gamesindustry.biz. Retrieved 2011-06-07.

[7] Tom Bramwell (2005-08-23). "It would be 'good to finish' BG&E - Michel Ancel". Eurogamer. Retrieved 2008-05-30.

[8] "Michel Ancel - next project is still a secret, Rayman platformer sequel possibility, and Jade talk". GoNintendo. 2007-11-03. Retrieved 2007-11-03.

[9] Jim Sterling (2008-05-07). "Who wants another Beyond Good & Evil rumor? I do I do I do!". Destructoid. Retrieved 2008-05-08.

[10] Régis Déprez (2012-07-01). "OPM : Ancel parle de Beyond Good & Evil 2" (in French). Gamekyo.com. Retrieved 2012-07-01.

[11] Boyes, Emma (2008-05-15). "Report: Beyond Good & Evil 2 on the way". GameSpot. Retrieved 2008-05-28.

[12] Brendan Sinclair (2008-05-28). "Beyond Good & Evil returns". GameSpot. Retrieved 2008-05-28.

[13] Matt Wales (2008-05-28). "Ubisoft Announces Beyond Good & Evil Follow-up". IGN UK. Retrieved 2008-05-30.

[14] Luke Plunkett (2008-05-15). "Beyond Good & Evil 2 Is For PS3, 360". Kotaku, Gawker Media. Retrieved 2008-05-28.

[15] "'Beyond Good & Evil 2' Already 1.5 Years into Development". WorthPlaying. 2008-12-19. Retrieved 2008-12-23.

[16] "Interview Michel Ancel - 21 mars 2009". Pro-gamers.fr. 2008-03-21. Retrieved 2009-11-08.

[17] Mike Fahey (2009-05-08). "Is This New Beyond Good & Evil 2 Footage?". Kotaku. Retrieved 2009-05-09.

[18] Laurent Detoc (2009-07-22). *Interview: Ubisoft's Detoc on Turning 'Gamers' into 'Players'* (Transcript). Interview with James Brightman. Industrygamers. Retrieved 2009-07-27.

[19] "Rayman Origins Video - E3 2011: History of Rayman Interview". GameTrailers. 2011-06-07. Retrieved 2014-06-02.

[20] "Beyond Good & Evil 2 is on hold". Gamezine. 2009-08-24. Retrieved 2009-08-24.

[21] "Ubisoft - Beyond Good and Evil 2 continuing on". Go-Nintendo. 2009-08-28. Retrieved 2009-08-28.

[22] "Ubisoft confirms Beyond Good and Evil 2 placed on hold at Gamescon (Germany)". *Ubisoft official forums*. 2009-09-01. Retrieved 2009-11-08.

[23] "EXCLU : Quel avenir pour Beyond Good & Evil 2 ?" (in French). Gameblog.fr. 2009-11-06. Retrieved 2009-11-29.

[24] Jim Reilly (2010-01-14). "Beyond Good & Evil 2 Still In Development". IGN. Retrieved 2010-06-08.

[25] "Beyond Good & Evil 2 Must Be Perfect". Kotaku, Gawker Media. 2010-08-19. Retrieved 2010-08-19.

[26] ""Keep the faith" for Beyond Good & Evil 2". Eurogamer.net. 2011-02-24. Retrieved 2011-05-18.

[27] Tom Phillips. "Eurogamer - Beyond Good & Evil 2 in "active creation"". Eurogamer.com. Retrieved 2012-09-19.

[28] Jeffrey Matulef. "Michel Ancel starts new studio, but remains at Ubisoft, who confirms that he's working on Beyond Good & Evil 2". Eurogamer.com.

[29] Val. "Beyond Good & Evil 2 existe : Nos infos exclusives !". PS4France.

[30] Percevault, Vincent (2008-05-29). "Christophe Heral confirmed to compose BGE2". Soundakt. Retrieved 2008-06-01.

[31] Alec Meer (2008-05-15). "Beyond Beyond Good & Evil". rockpapershotgun.com. Retrieved 2009-11-08.

[32] Kris Graft (2008-05-29). "Ubi: Beyond Good & Evil 2 More "Casual"". Next-Gen. Retrieved 2008-05-30.

[33] Matt Martin (2008-06-02). "Ubisoft's Yves Guillemot". gamesindustry.biz. Retrieved 2009-11-08.

[34] Rob Purchese (2009-04-03). "BG&E2 inspired by September 11". Eurogamer. Retrieved 2010-05-29.

[35] Rob Purchese (2009-04-03). "BG&E2 to be "more immersive, complex"". Eurogamer. Retrieved 2010-05-29.

2.3.4 External links

- Official teaser on Machinima's YouTube Channel

2.4 Black Mesa (video game)

Black Mesa (originally **Black Mesa: Source**; stylized as **BLλCK MESA**) is a third-party total remake of *Half-Life*, Valve Corporation's critically acclaimed debut title, on Valve's Source engine. The 40-person volunteer development team says they hope to create a more engrossing in-game world with more varied, complex environments; and more challenging, realistic gameplay.

During its eight-year development period, *Black Mesa* has been featured in several video game publications and received direct attention from Valve. Due to its long development time, the modification became notable for its delays, and dwindling updates on the status of its completion. The delays led to *Wired* awarding *Black Mesa* high spots on their "Vaporware Of The Year" lists in 2009 and 2010.[*][4][*][5]

The first part of *Black Mesa*, which included remakes of chapters "Black Mesa Inbound" to "Lambda Core", was released as a standalone download on September 14, 2012.[*][6][*][7] Valve, through public voting on the Steam Greenlight program, approved *Black Mesa* for distribution on Steam.[*][8][*][9] On May 5, 2015, *Black Mesa* was released as early access on Steam.

2.4.1 Gameplay

See also: Gameplay of Half-Life

Black Mesa is a first-person shooter that requires the player to perform combat tasks and puzzle solving to advance through the game. The core gameplay remains unchanged from the original *Half-Life*; the player can carry a number of weapons that they find through the course of the game, though they must also locate ammunition for most weapons. The player's character is protected by a hazard suit that monitors the player's health and can be charged as a shield, absorbing a limited amount of damage. Health and battery packs can be found scattered through the game, as well as stations that can recharge both.

2.4.2 Plot

See also: Story of Half-Life (video game)

The plot of *Black Mesa* is almost identical to *Half-Life's* storyline, playable through the "Lambda Core" chapter.[*][10] As in the original game, the player controls Gordon Freeman, a scientist working at the Black Mesa Research Facility. He is tasked to place a sample of a strange material into an electromagnetic instrument, using the Hazardous Environment Suit Mark IV to do so safely. However, the sample material causes a "resonance cascade", devastating the facility and creating an interdimensional rift to an alien dimension called Xen, bringing its alien creatures to Earth. Freeman survives, finds other survivors, and makes his way to the surface with the protection of his hazard suit to get help. Upon reaching the surface, however, he finds that the facility is being cleansed of any living thing - human or alien - by armed forces. From other scientists, Freeman finds the only way to stop the alien invasion is to cross over to Xen and destroy the entity holding the portal open.

2.4.3   Development

The "Surface Tension" chapter as it appears in *Half-Life*

The same scene, as seen in a development version of *Black Mesa*

With the release of *Half-Life 2* in 2004, Valve Corporation re-released several of its previous titles, ported to their new Source game engine, including the critically acclaimed 1998 game *Half-Life* named *Half-Life: Source*. The Source engine is graphically more advanced than the GoldSrc engine used for the original versions. *Half-Life: Source* features the Havok physics engine and improved effects for water and lighting. The level architecture, textures, and models of the game however, remained unchanged.

Half-Life: Source was met with mixed reviews. IGN liked the new user interface and other technical features but noted that it did not receive as many improvements as Valve's other Source engine ports.[11] GameSpy said that while it was a "fun little bonus", it was "certainly not the major graphical upgrade some people thought it might be".[12] Valve CEO Gabe Newell is quoted as saying that a complete remake of *Half-Life* by fans of the game using Source was "not only possible···but inevitable".[13]

Black Mesa began as the combination of two independent volunteer projects, each aiming to do just that: completely recreate *Half-Life* using Source. The *Leakfree* modification was announced in September 2004. *Half-Life: Source Overhaul Project* was announced one month later. After realizing their similar goals, project leaders for both teams decided to combine efforts; they formed a new 13-person team titled *Black Mesa: Source*.[13] The "Source" in the project's title was later dropped when Valve asked the team to remove it in order to "stem confusion over whether or not [it was] an endorsed or official product", which it at the time was not.[14]

The team now consists of 40 volunteer level designers, programmers, modelers, texture artists, animators, sound engineers, voice actors, and support staff.[15] They have stated they want *Black Mesa* to be similar to *Half-Life* in gameplay and story, but changes will be made to take advantage of Source's more advanced features. Changes to the story will not divert from, or alter, the overall storyline of the *Half-Life* series.[13] Level designers have shortened or modified some areas of the game that "didn't make any sense", or were "tedious" in the original. Maps will also be of a larger scale, for instance the hydro-electric dam, which is now "twenty or thirty times" larger.[16]

Originally based on the version of Source released with *Counter-Strike: Source* in 2004, the project switched to a more recent version released with Valve's *The Orange Box* in 2007. This new version included more advanced particle effects, hardware-accelerated facial animation, and support for multi-core processor rendering among other improvements.[17][18][19] The team recently stated that they have moved *Black Mesa* to Valve's new 2013 version of Source, with faster load times and Mac OS X and Linux support.[20]

2.4.4   Marketing and release

The developers released a teaser trailer in 2005, and a full-length preview trailer in 2008. They also released images, videos, and concept art during the project's development. *Black Mesa* was given an official release date of "late 2009" in the spring of 2009, but this date was changed to "when it's done" after the development team was unable to fulfill this date.[16]

On June 10, 2012, the *Black Mesa* development team announced that new "media" would be released once their Facebook page reached 20,000 likes.[21] This goal was reached on June 11, 2012 when 8 new screenshots were released, along with an announcement of the start of a "social-media campaign" towards their first release.[22]

On September 2, 2012, project leader Carlos "cman2k"

Montero announced that the first *Black Mesa* release would take place on September 14, 2012.[23][24][25]

The first section of *Black Mesa* was released on September 14, 2012,[6][26] distributed as a free download.[27] It has been confirmed that *Black Mesa* will also be distributed via Steam; the remake was among the first ten titles whose release on the platform was approved using Valve's crowdvoting service Steam Greenlight.[8] The initial release consists of remakes of all *Half-Life*'s chapters except those set on the alien world "Xen", which the team intends to expand for inclusion in a future release.[25] The development team estimates that the initial release of *Black Mesa* gives players eight to ten hours of content to complete.[25] In November 2013, the team confirmed they have been given the go-ahead from Valve to release a commercial version of the *Black Mesa* product via Steam. The team plans to use the additional funds to improve the game's implementation in Source and to include additional features; the free version will continue to be offered.[28]

On May 5, 2015, another trailer was released on the Steam website. The game was also released as a stand-alone game on Steam's Early Access.[29]

2.4.5 Soundtrack

In addition to the modification itself, the game's thematic score, produced by sound designer Joel Nielsen, has been independently released as a soundtrack.[30]

2.4.6 Modifications

In *Surface Tension Uncut*, an unofficial mod for Black Mesa, the "Surface Tension" chapter was expanded to include certain areas of the chapter from the original game that weren't released with the remake as a result of the developer working on that part leaving before his work was finished.[31] The developer, Chon Kemp, known on the *Black Mesa: Community Forums* by the pseudonym TextFAMGUY1,[32] is currently working on modifying the "On a Rail" chapter, which is in Beta stage.[33]

2.4.7 Reception

Prior to release

During its development, *Black Mesa* has received attention from several video game publications. It has been featured in articles from *Computer Gaming World*, *PC PowerPlay*, and *PC Gamer UK* magazines. Valve published a news update about the modification on their Steam digital distribution platform in 2007 saying that "We're as eager to play

[*Black Mesa*] here as everyone else." [34]

The project was awarded *Top Unreleased Mod* by video game modification website Mod DB in 2005 and 2006.[35][36] Mod DB gave the project an honorable mention in their choice of *Top Unreleased Mod* in 2007.[37]

After receiving a development version of *Black Mesa* in December 2009, *PC PowerPlay* magazine said that the game's setting "looks, sounds, [and] plays better than ever before". The "subtle" changes from the original *Half-Life* were said to have a "substantial" overall impact. They also noted the project's "frustrating" then-five-year development time, and current lack of release date, but added that the developers were making progress.[16]

Initial release

After the mod was released, early impressions of the game were very positive,[43] receiving a score of 86/100 on Metacritic, based on nine reviews.[38] The game was praised for its high polish, with many critics comparing its quality to that of an official Valve Corporation title.[42][41][39] Destructoid praised the game for the improvements it made over the original *Half-Life*, saying it was "something that felt very familiar, [but also] very fresh." [40]

Black Mesa won ModDB's Mod of the Year Award for 2012.[44][45]

2.4.8 References

[1] "Black Mesa: Re-visit the world that started the Half-Life continuum". *Black Mesa*. Retrieved February 25, 2013.

[2] "Black Mesa Community Forums". James Kane. Retrieved November 1, 2011.

[3] "Black Mesa Community Forums". Carlos "cman2k" Montero. Retrieved September 1, 2012.

[4] Calore, Michael (December 21, 2009). "Vaporware 2009: Inhale the Fail". *Wired (magazine)*. Condé Nast Publications. Retrieved September 22, 2012.

[5] Calore, Michael (January 3, 2011). "Vaporware 2010: The Great White Duke". *Wired (magazine)* (Condé Nast Publications). Retrieved September 15, 2012.

[6] Hamilton, Kirk (September 14, 2012). "You Can Play Black Mesa Right Now". *Kotaku*. Retrieved September 15, 2012.

[7] Hubi, Josh (September 5, 2012). "Launch Details (Release Personnel)". *Black Mesa: Community Forums*. Retrieved September 11, 2012. At launch, a launch pad will be available offering multiple mirrors and options for download.

[8] "First Titles Get The Community's Greenlight". *Steam Community*. Valve Corporation. September 11, 2012. Retrieved September 22, 2012.

[9] Schreier, Jason (September 11, 2012). "Black Mesa, Nine Other Games Get Greenlit On Steam". *Kotaku*. Gawker Media. Retrieved September 15, 2012.

[10] "Fans finally finish Half-Life remake Black Mesa". *Metro News*. September 4, 2012. Retrieved March 30, 2014.

[11] McNamara, Tom (November 18, 2004). "Half-Life: Source: What's the big hoo-ha?". IGN. Retrieved March 13, 2010.

[12] Accardo, Sal (November 17, 2004). "Half-Life: Source". GameSpy. Retrieved March 13, 2010.

[13] Elliot, Shawn. "Black Mesa: Source" (PDF). *Computer Gaming World* (Cambridge, MA: ZDNet) (257). ISSN 0744-6667. Retrieved March 13, 2010.

[14] Hill, Jason (February 16, 2009). "Your Turn: Returning to the Source". The Age. Archived from the original on March 30, 2010. Retrieved March 13, 2010.

[15] "Team Members". Black Mesa Modification Team. January 4, 2010. Retrieved March 13, 2010.

[16] Kim, Paul (December 7, 2009). "Black Mesa: Why The Future of Half-Life 2... Is In The Past". *PC PowerPlay* (Macclesfield: Media House) (174). ISSN 1362-2722. Retrieved March 13, 2010.

[17] "Source - Rendering System". Valve. Retrieved August 8, 2009.

[18] "Face-to-Face with TF2's Heavy". *Steam*. Valve Corporation. May 15, 2007. Retrieved May 23, 2010.

[19] "Interview: Gabe Newell". PC Zone. September 11, 2006. Retrieved September 20, 2006.

[20] "Why Black Mesa doesn't support Mac/Linux Platforms". September 13, 2013. Retrieved January 19, 2014.

[21] Crossley, Rob (June 10, 2012). "Black Mesa returns to the surface - new 'media' Inbound". *ComputerAndVideoGames.com*. Future Publishing Ltd. Retrieved September 15, 2012.

[22] Crossley, Rob (June 11, 2012). "New Black Mesa screenshots begin final release phase". *ComputerAndVideoGames.com*. Retrieved June 11, 2012.

[23] "Community Update". *blackmesasource.com*. September 2, 2012. Retrieved September 2, 2012.

[24] Plunkett, Luke (September 2, 2012). "Black Mesa Mod Out in Two Weeks". *Kotaku*. Gawker Media. Retrieved September 15, 2012.

[25] "Fans resurrect Half-Life video game". *BBC News* (BBC). September 3, 2012. Retrieved September 3, 2012.

[26] Cobbett, Richard (September 14, 2012). "Black Mesa Source released – download it now!". *PC Gamer*. Future Publishing Ltd. Retrieved September 15, 2012.

[27] "Black Mesa FAQ". Black Mesa Development Team. Retrieved November 4, 2010.

[28] Matulef, Jeffrey (November 19, 2013). "Valve gives Black Mesa permission to be a commercial product". Eurogamer. Retrieved November 19, 2013.

[29] "Black Mesa". *Steam*. Crowbar Collective. May 5, 2015. Retrieved May 6, 2015.

[30] Nielsen, Joel. "Black Mesa - Soundtrack". *Black Mesa*. Retrieved September 2, 2012.

[31] Pearson, Craig (January 9, 2013). "Mod A Mod: Surface Tension Uncut Beefs Up Black Mesa". *Rock, Paper, Shotgun*. Retrieved January 13, 2013.

[32] Gera, Emily (January 11, 2013). "Half-Life mod Black Mesa receives restored 'Surface Tension' section in new meta-mod". *Polygon*. Retrieved January 13, 2013.

[33] Petitte, Omri (January 10, 2013). "Surface Tension Uncut mod restores missing Black Mesa level content". *PC Gamer*. Retrieved January 13, 2013.

[34] "Friday, January 26, 2007". *Steam*. Valve Corporation. January 26, 2007. Retrieved May 23, 2010. *Congratulations to the Black Mesa for Half-Life 2 MOD team for picking up the Most Anticipated MOD Award for the coming year from Mod DB. Over 80,000 votes were cast for MODs built for a number of different games, and they have been crowned this year's most wanted. [...] We're as eager to play it here as everyone else.*

[35] Reismanis, Scott (January 26, 2006). "Mods of 2005". *Mod DB*. DesuraNET Pty Ltd. Retrieved May 23, 2010.

[36] Reismanis, Scott (January 13, 2007). "Mods of 2006 - Player's Choice". *Mod DB*. DesuraNET Pty Ltd. Retrieved May 23, 2010.

[37] stenchy (January 28, 2008). "2007 Mod of the Year Awards - Player's Choice Winners Showcase". *Mod DB*. DesuraNET Pty Ltd. Retrieved May 23, 2010. *An honorable mention is due to the mod Black Mesa which continues to poll extremely strongly year after year, but misses out on a place in the top 5 because you cannot win a spot in the best unreleased category twice.*

[38] "Black Mesa". *Metacritic*. Retrieved March 30, 2014.

[39] Kelly, Andy (September 17, 2012). "Black Mesa Review: Half-Life Still Packs A Punch In 2012 (With Help From Some Modders)". *Computer and Video Games*. Retrieved March 30, 2014.

[40] Derocher, Joshua (September 23, 2012). "Review: Black Mesa". *Destructoid*. Retrieved March 30, 2014.

[41] Ambròs, Albert (November 4, 2012). "Análisis de Black Mesa" [Review of Black Mesa]. *Eurogamer* (in Spanish). Retrieved March 30, 2014.

[42] "Black Mesa review". *GamesTM*. October 12, 2012. Retrieved March 30, 2014.

[43] "Impressions: Black Mesa is awesome". *Destructoid*. September 15, 2012. Retrieved September 16, 2012.

[44] "Mod of the Year 2012". *Mod DB*. DesuraNET Pty Ltd. December 25, 2012. Retrieved March 30, 2014.

[45] "Mod of the Year 2012 results video".

2.4.9 External links

- *Black Mesa: Source* Official Website

2.5 Bob's Game

Bob's Game was a role-playing video game being developed by independent software developer Robert Pelloni. The project is most notable for Pelloni developing the game using open source software development tools and Nintendo's refusal to license him the official SDK.

2.5.1 Development

Bob's Game was to be a 2D role-playing video game developed solely by Pelloni. According to an interview he gave to the *Orlando Sentinel*, Pelloni spent over five years and over 15,000 hours working on the game.*[1] The game was to feature over 200 characters and, according to the interview, "more gameplay than just about anything out there on the portable system".*[2] In August 2008, Pelloni posted a preview of the game on YouTube that had received over 100,000 views by September 15, 2008.*[2]

Development started as a result of a conversation about video games that Pelloni had with friends at a restaurant. He discussed creating a video game based in the suburbs that had a Dungeons & Dragons mindset, similar to the *EarthBound* series. Pelloni first got the idea of creating *Bob's Game* from watching his brother play *Dragon Warrior*, whose story, he said, he found confusing. He also drew motivation from other titles such as *Super Mario 64*, *Super Metroid*, the *Dance Dance Revolution* series, and a similar game developed by one person titled *Cave Story*. He started to brainstorm ideas for this game, scribbling notes on a napkin. According to the *Sentinel* interview, Pelloni

was fascinated by other similar video game projects getting published after participating in online Internet forums. At the same time, he was also "disheartened" over how the video game industry did business, saying that "it's a standard practice for some publishers to take a game engine and put in licensed assets to coincide with, say, a movie release for example".*[2] This motivated Pelloni to self-develop *Bob's Game* and, as a result, he spent most of 2006 and 2007 in isolation while developing the game.*[2]

Pelloni said that the hardest part of developing the game was the background graphics, which he drew by himself despite having no artistic experience. He said that he most enjoyed writing the dialogue and designing the gameplay elements. He would not consider the game "100% complete" until he received the software development kit (SDK) from Nintendo, which would allow him to compile the game according to Nintendo's code specifications. However, he stated in the *Sentinel* interview, "[Y]ou can't get access to [the SDK] unless you've published a game before".*[2] Pelloni had received responses from some video game publishers but did not start talks with any of them as he wanted to retain creative rights to the game. As a replacement for the standard credits roll at the end of most video games, he said he would provide a summary of the making of the game.*[2]

2.5.2 Rejection and protest

According to an article in *The Escapist*, Pelloni was directed by Nintendo to talk to the *Wario World* division, which directed him to marketing; marketing then directed him back to the *Wario World* division.*[3] Nintendo told him that they would inform him of their decision whether to grant him an SDK for the game in between six and eight weeks. No response came from Nintendo. After 17 weeks of trying and failing to get Nintendo to provide him with the SDK, on December 11, 2008, Pelloni decided to publicly protest by locking himself in his room for 100 days or until Nintendo provided him with the SDK, whichever came first.*[1] According to Owen Good of *Kotaku*, he staged the protest in an effort to gain publicity by making Nintendo look like a corporate bully beating down on an indie game developer.*[4] His room had no Internet access (save for broadcasting a live feed of him in his room via a webcam) or television and had only a mobile phone with which he could make calls and send emails and the materials he needed to work on the game.*[5] Pelloni made the following comment when he decided to protest:*[3]

> "I cannot leave this viridian room. The door is locked and barricaded from the outside. I am sleeping behind the camera, and yes- I've got a shower. Food is delivered once a week by a

friend...This is my 100 day protest to Nintendo!"
Robert Pelloni

Pelloni's protest garnered popularity on various Internet forums and websites.[6] On the 21st day, December 31, 2008, he started to release addresses of Nintendo executives and to send Nintendo of America president Reggie Fils-Aime holiday greetings. He also threatened to bundle the game as a killer app using a Nintendo DSi homebrew device, and said that he was translating the game into several languages, planning to release it worldwide to "significantly cut into Nintendo's bottom line".[7] According to Chris Greenhough of *Joystiq*, he said that he was talking to a Chinese firm about releasing *Bob's Game* on a flash cartridge and negotiating with Walmart to distribute the game.[8] He also threatened to get the game released on other distribution platforms including Xbox Live Arcade, Steam, the iPhone, and the PlayStation Network.[9] On January 6 he declared that he was better than Shigeru Miyamoto, Hiroyuki Ito, Hideo Kojima, Adrian Carmack, and Will Wright combined.[10] Though he apologized for his comment two days later, according to Andy Chalk in an *Escapist* article, he threatened to "exact a horrific vengeance if the company continued to deny him the SDK". At this point, he started to complain about paranoia and having persistent headaches.[10]

2.5.3   Protest aftermath

The protest ended on the 30th day on January 10, 2009, with Pelloni saying that he was suffering from a "wicked headache". Pelloni described Nintendo as a "heartless corporation, only interested in the biggest profits".[11] He ransacked his room in frustration and posted a lengthy comment on his website, declaring his defeat. In light of the statement and the lack of any visible movement on the webcam,[4][9] a concerned user from the /v/ board on 4chan was able to retrieve Pelloni's telephone number from his Whois information. The user contacted his sister in an attempt to have somebody check on him.[4] Then, on January 11, the police broke into his room to check on him, finding him okay.[9]

According to Rob Hearn in an article on the website *Pocket Gamer*, shortly after conceding, Pelloni resumed his protest and attack on Nintendo. He had apparently worked himself into *Bob's Game*, portraying himself as the game's main antagonist and final boss. He redefined the object of the game as being to take down "Gantendo" (a portmanteau of "Ganon" and "Nintendo"), with the game's main protagonist being named "Yuu". In addition, he had retooled the story to incorporate events leading up to and after his protest.[12] On February 1, Griffen McElroy of *Joystiq*

reported that Pelloni had vandalized the Nintendo World Store in New York City, saying that this was "Level 50" of his newly revived protest.[13]

On February 6, 25 weeks after sending his request for an SDK, Pelloni received a letter from Nintendo rejecting his request. According to JC Fletcher of *Joystiq*, this was the same form letter that Xiotex Studios received on their rejection to develop games for WiiWare. In the letter, Nintendo said that they require "secure business facilities, sufficient equipment and staffing, financial stability and other attributes that would distinguish the developer" and that they deal with confidential information, making them highly selective in whom they grant SDK access.[14]

In March 2009, Pelloni announced on his website that the protest and ensuing events had been a viral marketing ploy to advertise *Bob's Game*. According to an email he sent to the press, he said that he had been able to fool the entire Internet gaming community and expressed disappointment in the media, saying that "'angry developer litters' is considered more newsworthy than 'angry developer'".[15] He referred to the marketing of the game as "an old-school marketing style for an old-school style game".[15] According to Jim Sterling of *Destructoid*, he claimed that his stunts "are the mark of somebody who deserves to be a part of the game industry".[15]

On March 31, it was reported that Pelloni had released a playable demo of *Bob's Game*, which, according to his website, required a flash cartridge and was playable on the NO$GBA emulator.[16] On April 2, 2009, MTV.com's Stephen Totilo interviewed Reggie Fils-Aime about the *Bob's Game* incident. Fils-Aime said that Pelloni had applied to be a licensed developer for Nintendo, but Nintendo rejected his application after a standard evaluation because he did not meet their requirements. He noted that he is happy that people are motivated by games developed under similar circumstances such as 2D Boy's *World of Goo* as well as *Tetris* and *Pokémon*, that Nintendo enjoys "taking big ideas with small budgets and bringing them to life".[6]

On March 4, 2011, Pelloni announced a new portable console called the nD that would be sold for $20 with *Bob's Game* as the first title. On June 9, 2011, the final day of Electronic Entertainment Expo 2011, Pelloni uploaded a video called "nD Commercial" to YouTube. In early 2013 Bob silently deleted all the references to the nD and put up a live stream of him working on to his official site bobs-game.com. As of May 8, 2013, the website has been updated to show a "new phase" of *Bob's Game* called "Bob's Game Online: nDworld". After playing a short demo, players are asked to register and after doing so are directed to a website telling them to wait for future updates; at first they were also told they would have to pay for the service in the future. On November 25, 2013, Pelloni made a Kickstarter for the puzzle game from *Bob's Game*, which failed on December 15, 2013, with $477 out of the goal of $6,667. Despite the Kickstarter failing, the game was released on the Ouya on January 1, 2014. Shortly after the Kickstarter failed, Pelloni created a Patreon page, which he later removed. On April 23, 2014, Pelloni launched the first and only Kickstarter for the previously-announced action-RPG *Bob's Game*, saying that if the crowd-funding venture was successful, he would invest the money to work from a "hack-van" in order to complete development on the title, and if not, he would put development of the game on indefinite hiatus. On May 22, 2014, with 17 hours left, the game was successfully funded.[17] Pelloni acknowledged this success shortly thereafter on his website, but as of July 1, 2014 he has not posted any further updates. According to Forbes, the developer has given up on the project, and plans to refund backers.[18]

2.5.4 References

[1] Davis, Ryan (2009-01-17). "Bob's Game Gets Burgled". Giant Bomb. Retrieved 2009-08-10.

[2] Simantov, Matthew (2008-09-15). "Interview with the creator of Bob's Game - (probably) the biggest game ever created by 1 person". *Orlando Sentinel* (blog). Retrieved 2009-08-10.

[3] Nq, Keane (2008-12-22). "Bob's Game Developer Stages 100 Day Protest to Nintendo". The Escapist. Retrieved 2009-08-10.

[4] Good, Owen (2009-01-10). "[Updated] Bob's Protest and Bob's Game is Over". *Kotaku*. Retrieved 2009-08-10.

[5] McElroy, Justin (2008-12-22). "'Bob's Game' dev confines self in Nintendo protest". *Joystiq*. Retrieved 2009-08-10.

[6] Totilo, Stephen (2009-04-02). "Nintendo Finally Comments On 'Bob's Game' Situation". MTV.com. Retrieved 2009-08-10.

[7] McElroy, Justin (2008-12-31). "Bobwatch Day 21: Things get kind of weird". *Joystiq*. Retrieved 2009-08-10.

[8] Greenough, Chris (2009-01-05). "Bob's Saga rumbles on". *Joystiq*. Retrieved 2009-08-10.

[9] Yerby, Anthony (2009-01-17). ""Bob's Game" creator is officially out of his mind". *Aeropause*. Archived from the original on 2009-01-22. Retrieved 2009-08-10.

[10] Chalk, Andy (2009-01-12). "Bob's Game Guy Gives Up". The Escapist. Retrieved 2009-08-10.

[11] McElroy, Justin (2009-01-10). "'Bob's Game' 100 day sit-in protest ends early, disturbingly". Joystiq. Retrieved 2009-08-10.

[12] Hearn, Rob (2009-01-13). "Bob's Game creator Bob Pelloni's 100 day protest is back on". *Pocket Gamer*. Retrieved 2009-08-10.

[13] McElroy, Griffin (2009-02-01). "Jilted Bob's Game creator fights back by littering". *Joystiq*. Retrieved 2009-08-10.

[14] Fletcher, JC (2009-02-06). "Nintendo denies official DS developer status to 'Bob's Game' creator". Joystiq. Retrieved 2009-08-10.

[15] Sterling, Jim (2009-03-15). "Bob lets the cat out of the bag, explains viral campaign". *Destructoid*. Retrieved 2009-08-22.

[16] Fletcher, JC (2009-03-31). "Bob's (playable) Game: Homebrew demo released". *Joystiq*. Retrieved 2009-08-10.

[17] https://www.kickstarter.com/projects/bobsgame/bobs-game

[18] http://www.forbes.com/sites/michaelthomsen/2015/01/14/bobs-game-kickstarter-to-be-refunded-after-developer-moves-on/

2.5.5 External links

- Official website of *Bob's Game*

2.6 CherryOS

The CherryOS logo

CherryOS was a PearPC based PowerPC G4 processor emulator for x86 Microsoft Windows platforms, announced on October 12, 2004. It was released to the public on March 9, 2005 and development stopped on May 6, 2005.[1] The program was called vaporware or a hoax, with critics calling the program's existence into question, because of numerous missed deadlines and failure to produce demonstration versions. This ceased when Slashdot announced that CherryOS brand software was available for download. Sometimes it disappeared as a downloadable option, and was re-issued with new file sizes and minor changes, perhaps to appease critics.

CherryOS was marketed by Maui X-Stream. The chief architect was said to be Arben Kryeziu. According to Maui X-Stream, it allowed users to install and run versions of Mac OS X on platforms that would not normally support such software, such as Pentium processor-based systems. It was advertised as being able to reach emulation speeds equalling roughly 80% of the system's total processor speed.

As of 2009, CherryOS and its website, CherryOS.com, are defunct.

The owner of Maui X-Stream, Arben Kryeziu, later went on to found Code Rebel.

2.6.1   License issues

Critics have alleged that CherryOS is simply a modification of PearPC.*[2] Since PearPC was released under the GNU General Public License (GPL) and CherryOS was not, this would have been a violation of the GPL.

CherryOS was re-released on 10 March 2005, with an easy-to-use graphical installation program a feature that PearPC lacks.

CherryOS was also alleged to have copied code from OpenVPN (using it as their network driver) and HFVExplorer (which was used as the "Drag 'n' Drop" interface).

2.6.2   Legal threats

In early 2005, college student Kristian Hermansen posted a very rudimentary analysis of CherryOS using IDA Pro software, which easily shows evidence of code reuse from many GPL projects. This finding caused him to independently investigate the alleged "author" of the code, Arben Kryeziu. In a brief long distance phone call to Hawaii with the author, under the guise of an interested corporate purchaser, it was immediately apparent that his stories didn't match up to what was being fed to the public media. This phone conversation was recorded, but the audio was never released.

In order to solidify the claims, Kristian privately consulted the help of Halvar Flake, a widely respected software reverse engineer. Using a custom tool called BinDiff, Halvar conservatively estimated that over 600 functions in CherryOS were nearly identical or were identical to that of a similar PearPC build.

In retaliation for illegal GPL code reuse, Kristian posted a complete binary of the latest CherryOS installer along with a valid key generator on his website. This bold action, he hoped, would stifle sales of CherryOS and force the parent company, Maui X-Stream Inc, to stop selling such software. However, when Maui X-Stream finally realized their software was being made publicly available to all, Kristian was served with a cease and desist order via email from Maui's lawyers, demanding the software be taken down. He refused to remove the software, and instead sought legal advice from the EFF. A short time later, Kristian was contacted by Eben Moglen via telephone, and officially retained him as a lawyer at no charge.

In an ironic twist to the story, the CherryOS lawyers threatened to take action against Kristian in his home state of Massachusetts. In the end, Maui X-Stream did not seek action against Kristian, and the CherryOS software was finally laid to rest in April 2005. Kristian removed the CherryOS software from his website a short time later, and instead started to investigate a completely different Maui X-Stream product, VX30. He has also shown, with help from others, that VX30 contains GPL code as well and should be further analyzed.

2.6.3   Aftermath

Since the Software Freedom Law Center would not do actions pro-bono, if made in Hawaii, the PearPC team began collecting donations to prepare for a lawsuit against Maui X-Stream and CherryOS.*[3] As CherryOS withdrew, this money became largely unnecessary, and was later donated to the Software Freedom Law Center, who represented individual developers of PearPC.*[4]

On April 5, 2005, the main page of CherryOS was changed to a simple message that said, "CherryOS is On Hold - until further notice." .*[5] A day later, the message was changed again to claim that CherryOS, "Due to Overwhelming Demand" , has open sourced itself.*[6]*[7] A re-release was purported to be coming in May according to the site.

Critics raised at the time issues about Maui X-Stream's use of the term "open source" .*[8] Jim Kartes, president of Maui X-Stream Inc., said to eWeek that the company "will be charging $14.95 to cover our cost of development and continuing development as well as other costs related to the marketing of the product. Whatever the buyer does with the code is their business as long as they don't embed it into another commercial product." *[9] The position of the critics was that in part six of the Open Source Definition of the Open Source Initiative, "No Discrimination Against Fields of Endeavor" , there is no allowance for restrictions against commercial use. Part six of the Debian Free Software Guidelines also does not allow discrimination against fields of endeavor. The Free Software Foundation also says "A free program must be available for commercial use, commercial development, and commercial distribution."[10]

At one point Maui X-Stream had the GNU logo on the CherryOS homepage, but this has since been removed. More recently, the CherryOS.com website had been taken down reading "CherryOS is no more" with a leaf logo and links to "other emulators" (which did not include PearPC). As mentioned above, the page is now expired and run by a generic content provider.

2.6.4 References

[1] Arben Kryeziu's Blog

[2] Slashdot: CherryOS Not All It's Cracked Up To Be

[3] Slashdot: PearPC Trying to Sue CherryOS

[4] Software Freedom Law Center: PearPC Donates Legal Defense Fund to SFLC

[5] DrunkenBlog: CherryOS goes offline

[6] MacWorld: CherryOS goes open source

[7] EDN: CherryOS: Innovator or Copycat?

[8] DrunkenBlog: CherryOS and OSS

[9] eWeeK: CherryOS Opens Up Code to Doubters

[10] GNU's Free Software Definition

2.6.5 External links

- Wired Article on CherryOS
- The pits in CherryOS
- Deconstructing Maui X-Stream
- Analyzing the Legality of CherryOS

2.7 Cipher Complex

Cipher Complex is an unreleased stealth action video game that was being developed for the Microsoft Windows, PlayStation 3 and Xbox 360 systems by Edge of Reality.[2] The game was first announced in June 2006, by which time it was reported to already have been in production and self-funded for two years; the title was then signed with Sega in 2007,[3] but in July 2009 it was revealed that the game

had been cancelled by the publisher earlier that year after over half a decade of work, as noted in the LinkedIn profile of a former internal producer at the studio. Further details surrounding the cancellation and the game's fate in their entirety remain unclear to this day, as no official announcement regarding its status has been made by either Sega or Edge of Reality,[4] however, more recently, a U.S. federal trademark registration was filed on September 11, 2009, which as of October 19, 2011, has been granted a third extension. While confirming nothing else, this does indicate that Edge of Reality has retained the intellectual property rights to the game.[5]

Although the game was widely believed to be permanently cancelled by Sega, game designer Phil Fogerite (who worked as level designer on the project from March 2008 to April 2009) has most recently stated that Edge of Reality itself had in fact decided to put the game on indefinite hold.[1]

2.7.1 Gameplay

Edge of Reality was hoping to "revolutionize" the stealth action genre. The main character carries out various actions with speed, precision, strength and cunning stealth to neutralize enemies.

In June 2010, gameplay footage of a level from *Cipher Complex* was leaked online via Dailymotion, depicting the player character (Cipher) infiltrating a steel mill in China, under orders to extract a General who is being held hostage; in addition, short clips of gameplay can be seen on Edge of Reality's official homepage.[6][7] An "in-game screenshot" of the steel mill level, as well as some concept art, textures and logos from the game have also been posted to Picasa by the lead environment artist, who has since clarified that it is a demo level.[8][9] Part of an early cinematic as well as some character animations were again posted via Dailymotion in March 2011.[10]

2.7.2 Plot

The official storyline, as quoted from the game's original press release, is as follows:

> U.S. surveillance satellites detect activity onboard the decommissioned Soviet Bargration Missile Defense Station 4 off the east coast of Siberia. When the Russians deny the U.S. access to the facility, Department of Defense strategists suggest that a small, plausibly deniable reconnaissance mission be sent in to investigate. The Defense Threat Reduction Agency is given the go-ahead for operation BLACKOUT, the insertion

of a single expert Operator on Russian WMDs and launch facilities. Lt. Col. John Sullivan, callsign: Cipher is air dropped in, and what was supposed to be primarily a reconnaissance mission becomes a race against a terrorist threat; one with implications that will shake the foundations of American democracy and freedoms.

It is unknown to what extent the game's plot changed during development. As such, the synopsis outlined above may not be indicative of the final plot intended for the game.

2.7.3 References

[1] "Phil Fogerite (LinkedIn profile)". LinkedIn. Retrieved 17 December 2011.

[2] "Interview with Evan Bell". All About Coding. Retrieved 23 November 2010.

[3] "*Cipher Complex* is still in production". PSU.com. Retrieved 23 November 2010.

[4] "*Cipher Complex* canned?". GameSpot. Retrieved 23 November 2010.

[5] "*Cipher Complex* – Trademark by Edge of Reality, Ltd.". Trademarkia. Retrieved 6 April 2011.

[6] "*Cipher Complex* gameplay footage". Dailymotion. Retrieved 30 September 2010.

[7] "Edge of Reality company homepage". Retrieved 17 August 2011.

[8] Kohler, Rick. "*Cipher Complex* demo screenshot". Picasa. Retrieved 17 August 2011.

[9] Kohler, Rick. "Gallery". Picasa. Retrieved 17 August 2011.

[10] "*Cipher Complex* cinematics". Dailymotion. Retrieved 2 April 2011.

2.7.4 External links

- Announcement
- IGN
- Official website

2.8 Demons of Mercy

Demons of Mercy is an upcoming video game for the Xbox 360. *Demons of Mercy* will be developed by Maxum Games, and will be released alongside a comic-book miniseries created by R.H. Stavis. There has been very little new information of the game since its announcement.

2.8.1 References

- "Maxum Shows Mercy". IGN. 2006-08-08. Retrieved 2008-08-13.

2.8.2 External links

- IGN Demons of Mercy

2.9 Development hell

Development hell or **development limbo** is media industry jargon for a state during which a film or other project remains in development without progressing to production. A film, video game, television program, screenplay, computer program,[1] concept, or idea stranded in development hell takes an especially long time to start production, or never does. Projects in development hell are not officially cancelled, but work on them slows or stops.

2.9.1 Overview

Film industry companies often buy the film rights to many popular novels, video games, and comic books, but it may take years for such properties to be successfully brought to the cinema, and often with considerable changes to the plot, characters, and general tone. The original creators of the source material usually have very little to no involvement in the films' creative control, creating a divide among fans.[2] This pre-production process can last for months or years. More often than not, a project trapped in this state for a prolonged period of time will be abandoned by all interested parties or canceled outright. As Hollywood starts ten times as many projects as are those released, many scripts will end up in this limbo state.[3] This happens most often with projects that have multiple interpretations and affect several points of view.[4][5]

2.9.2 Causes

In the case of a film or television screenplay, the screenwriter may have successfully sold a screenplay to producers or studio executives, but then new executives assigned to the project may raise objections to prior decisions, mandating rewrites and recasting. As directors and actors join the project, further rewrites and recasting may be done, to accommodate the needs of the new talents involved in the project. Should the project fail to meet their needs, they might leave the project or simply refuse to complete it, resulting in further rewrites and recasting. At any point, a project may be forced to begin again from scratch.

It may also be the case that the screenwriters have an issue with the final rights agreement after signing an option, requiring research on the chain of title. The project may be stuck until the situation is resolved and project participants are happy with the full terms, or the project is abandoned.

When a film is in development but never receives the necessary production funds, another studio may execute a turnaround deal and produce the film to make it successful. An example of this is when Columbia Pictures developed, but then stopped production of *E.T. the Extra-Terrestrial*. Universal Pictures then picked up the film and made it a success. If a studio completely abandons a film project, the costs are written off as part of the studio's overhead.*[6] Sometimes studios or producers will deliberately halt production in order to stop competition on a different project, or to ensure that people invested will be available for other projects that the studio prefers.

During a potential writer's strike in 2001, major studios wanted to spend less time and energy bidding on longer-term developments, such as film rights to books. Instead they focused more on buying projects that would immediately receive a green-light such as big budget action thrillers, and high concept comedies written by established and credible writers. Studio executives put all uncertain scripts and pitches on the shelves during this time to avoid taking a chance on a long-term development, and only wanted projects that were ready to go into production. Some studios and producers still bought film rights to books, but only ones that had successful sales. Examples of this are Dino De Laurentiis' $9 million acquisition of Thomas Harris' *Hannibal* and Miramax purchasing Mario Puzo's *Omertà* for $2–$3 million.*[7]

The concept artist and illustrator Sylvain Despretz has suggested that "Development hell doesn't happen with no-name directors. It happens only with famous directors that a studio doesn't dare break up with. And that's how you end up for two years just, you know, polishing a turd. Until, finally, somebody walks away, at great cost." *[8]

2.9.3 Related concepts

In software development, unreleased products that have been in long-term development are considered a type of vaporware. In film and television screenplay, unreleased products that have been in long-term development are considered as "vaporfilm". The anime OVA adaptation of *Alien Nine* has been cited by fans and critics as an example of "vaporfilm" because it was put on hiatus in 2002 after four episodes.

2.9.4 Examples

Films

- ***Alien vs. Predator***
  Alien vs. Predator was first planned shortly after the 1990 release of *Predator 2*, to be released sometime in 1993. It was halted for more than a decade, with constant actor changes, restarts, and failed promotions of the film until it was finally released in 2004.*[9]

- ***Akira***
  Warner Bros. has been developing a live-action American version of the animated film for years. As of January 6, 2012, Warner has "shut down" production for the fourth time.*[10]

- ***Atlas Shrugged***
  Film and later television adaptations of Ayn Rand's novel were in development hell for nearly 40 years*[11] before the novel was finally brought to screen in the first part of a trilogy in 2011. *Part II* appeared in 2012 and *Part III*, was released in September 2014.*[12]

- ***Atuk***
  A film adaptation of the novel *The Incomparable Atuk*. Several principals involved in the film have died during the film's development time, now over a decade.*[13]

- ***Austin Powers 4***
  Austin Powers 4 was first announced by Mike Myers. "There is hope!". "We're all circling and talking to each other. I miss doing the characters." *[14] In July 2008, Mike Myers stated that he had begun writing *Austin Powers 4*, and that the plot is "really about Dr. Evil and his son." *[15] In September 2013, when asked about the future of *Austin Powers*, Myers answered "I'm still figuring that out." *[16]

- ***Beverly Hills Cop III***
  Went through multiple script revisions, including a treatment that had Axel Foley teaming up with a Scotland Yard detective to be played by Sean Connery until being finally released.*[17]*[18]*[19]*[20]

- The failure of ***Batman & Robin*** in 1997 also hindered many attempts to produce a fifth Batman movie until Warner Brothers opted to reboot the franchise in 2005, resulted on *Batman Begins* with far greater success.*[21]

- ***The Brazilian Job***
  A sequel to the 2003 remake of *The Italian Job* was in development by the summer of 2004, but has faced multiple delays. Principal photography was initially slated to begin in March 2005, with a projected release

date in November or December 2005.*[22] However, the script was never finalized, and the release date was pushed back to sometime in 2006,*[23] and later summer 2007.*[24] Writer David Twohy approached Paramount Pictures with an original screenplay entitled *The Wrecking Crew*, and though the studio reportedly liked the idea, they thought it would work better as a sequel to *The Italian Job*.*[25] Gray was slated to return as director, as well as most, if not all, of the original cast.*[24]*[25] At least two drafts of the script had been written by August 2007, but the project had not been greenlit.*[26]

- **Dallas Buyers Club**
  The screenplay was written in September 1992 by Craig Borten. Throughout the 1990s, he wrote 10 different scripts, hoping for it to be picked up. It was unable to secure financial backing, going through three different directors, finally being released in 2013, with Jean-Marc Vallée directing.*[27]

- **Foodfight!**
  In 2004, the CGI film *Foodfight* was announced. Described as "*Toy Story* in a supermarket", the film promised to bring together over 80 famous advertising characters with voice talent including Charlie Sheen, Hilary and Haylie Duff, Wayne Brady, and Eva Longoria. Director Lawrence Kasanoff expected it to be a commercial hit and merchandise for the movie appeared on store shelves before the film had a release date. The film ran into many problems.*[28] After several years, a trailer*[29] was finally shown at AHM in 2011, a company bought the DVD distribution rights for the film in Europe,*[30] and a quiet video-on-demand American release came in 2012, to extremely negative reviews and was a financial failure.

- **ID Forever Part I and II**
  The sequels to *Independence Day* were in development hell from 1997 until 2009, when director Roland Emmerich announced the pre-production of the films to be shot back-to-back.*[31] However, *ID Forever Part I* was renamed to *Independence Day: Resurgence* for the scheduled June 24, 2016 release.*[32]

- **The Jetsons Live-action film**
  A live-action adaptation of The Jetsons was first announced in late 1984 by Paramount Pictures. The film was to be executive produced by Gary Nardino and released in 1985, but failed to do so.*[33] In the late 1980s Universal Studios purchased the film rights for *The Flintstones* and *The Jetsons* from Hanna-Barbera Productions. The result was the animated film *Jetsons: The Movie*, which was released in 1990. In May 2007, director Robert Rodriguez entered talks with Universal Studios and Warner Bros. to film a live action film adaptation of *The Jetsons* for a potential 2009 theatrical release, having at the time discussed directing a film adaptation of *Land of the Lost* with Universal. Rodriguez was uncertain which project he would pursue next, though the latest script draft for *The Jetsons* by assigned writer Adam F. Goldberg was further along in development.*[34] The film was to be released in 2012. However, in early 2012, Warner Bros. Pictures delayed indefinitely the release of the film. Also in 2012, Warner Bros. hired the screenwriting duo Van Robichaux and Evan Susser to rewrite the script. Producer Denise Di Novi said in 2011 that Rodriguez was off the project as his vision for the movie "wasn't a mainstream studio version". Kanye West reported via Twitter in February 2012 that he was in talks to be creative director on 'The Jetsons'.*[35]

- **The Keith Moon Movie**
  A biopic of The Who drummer Keith Moon was first floated by The Who's singer Roger Daltrey in 1994. A competing movie by Keith Moon's personal manager, Peter "Dougal" Butler, produced by Robert DeNiro and written by Dick Clement and Ian La Frenais, was cancelled in 1998 after Daltrey had Pete Townshend deny the use of music by The Who.*[36] Since then, some major names have been attached to the movie (a script by Alex Cox*[37] among many written, and a starring role for Robbie Williams*[38] or Mike Myers*[39]) but no script has yet gotten Roger Daltrey's approval.*[40] As of 2013, the movie is attached to Exclusive Media and Da Vinci Media Ventures.*[41]

- **Love & Mercy**
  Named after the 1988 song, a biopic of the Beach Boys' Brian Wilson was proposed that year with William Hurt as Wilson. Discussions for a feature-length biopic continued over the decades, but production did not take off until 2011 with director Bill Pohlad and screenplay writer Oren Moverman at the helm. The film was eventually released in 2014 starring Paul Dano and John Cusack as Wilson in a dual role.*[42]

- **Me and My Shadow**
  A animated fantasy comedy film from DreamWorks Animation that would feature the studio's signature CG animation mixed with traditional hand-drawn animation. Was announced in December 2010 and slated for a release date in March 2013.*[43] It would then see two release date changes, first to November 2013*[44] and then to March 2014.*[45] In February of 2013, it was announced that the film had gone back into development with an unknown release date.*[46]

- **The Postman**
  Author David Brin described the ten-year effort to get

his novel produced as a film. Production began in 1987, but the final film was not released until 1997. In the process, the screenplay went through so many revisions that the shooting script only loosely resembled the book, and later writers "borrowed" elements from the book to improve the film. The film was a box-office bomb and was negatively reviewed.*[47]

- **The Thief and the Cobbler**
The Thief and the Cobbler is an animated movie originally created by Richard Williams. It was 28 years in production before the rights were sold to Warner Bros. The newly appointed director, Fred Calvert, released it as *The Princess and the Cobbler* adding dialog to some characters and 4 songs. In 1993, the rights were sold yet again to Miramax, which edited the movie further and renamed it *Arabian Knight*. When it was finally released to theaters, it was a box office bomb. The Miramax version was later released on VHS and DVD, again with poor sales.

- **Sin City: A Dame to Kill For**
Sin City 2, which was announced for a 2008 release, did not enter production until 2012,*[48] and was released in 2014.

- **Superman Lives**
The name given to a project begun by producer Jon Peters in 1993 as **Superman Reborn**. The proposed film would have followed the comic story line known as *The Death of Superman*. Jonathan Lemkin was hired to write the initial script, but Peters brought on a series of additional screenwriters to overhaul the script, including Gregory Poirier in 1995 and Kevin Smith in 1996. Director Tim Burton became attached to the film, with Nicolas Cage cast as the Man of Steel, and several more screenwriters were brought on board for several more rewrites. Burton backed out in late 1998 citing differences with producer Peters and the studios, additional writers and directors were attached to the project at various times over the next few years and instead to direct on *Sleepy Hollow*. Peters project went through several more permutations before evolving into *Superman Returns*, released in 2006, 13 years after initial development began.*[49]*[50]

- **Warcraft**
A live-action adaptation of the *Warcraft* series was first announced in 2006.*[51] The film spent several years in development hell before the project advanced. It is scheduled for a 2016 release.*[52]

- **The X-Files: I Want To Believe**
The second film based upon the popular American television show *The X-Files* began pre-production planning in 2001 and was announced for release in 2003 to follow the show's ninth season, but languished in development until it was finally produced for its release in the summer of 2008, six years after the television show had ended.*[53]*[54]

- **Mad Max: Fury Road**
The fourth film in the *Mad Max* franchise began development in the mid 90's, but struggled to find financing. Production was set to begin in 2001, but was halted due to the 9/11 attacks, director George Miller, instead choosing to focus on *Happy Feet*. After Mel Gibson lost interest, the role of Max was recast with Tom Hardy in the lead role. Production finally began in 2012, with reshoots in 2013, and the film was released in 2015 to critical acclaim.

- **Timeless**
Timeless is an original story written by Michael Bartlett (author of The Zombie Diaries), set in a dystopian future about a drug addicted time-traveling hit-man.*[55] There was talks of production in 2009, with the release of a poster and concept art.*[56] It was said that actors such as Jürgen Prochnow, Robert Kazinsky, Lee Ryan, James Fisher, were attached and Alexis Bledel was being rumored to be in the movie, and a rumored release date in 2011.*[57] There were however problems with finances as the production company was unwilling to commit 100%.*[58] Film production was completely haltered with the release of Looper, as Bartlett felt the two were too similar.*[59] As of 2013, Boundless Pictures optioned for the *Timeless* script,*[60] with a release date of 1. January 2016.*[61]

Music

- **The Smile Sessions**
Archival recordings of the Beach Boys unfinished album *Smile* took nearly 45 years to compile for a dedicated release. Numerous complications contributed to its excessively protracted delay, including bandleader Brian Wilson's irrational fear of the album. Brother and bandmate Carl Wilson compared the album's structuring to editing a film, as compiler Alan Boyd explains, "I think he was right about that. The kind of editing that the project required seemed more like the process of putting a film together than a pop record." *[62]

- **Chinese Democracy**
Rock band Guns N' Roses began work on this album in the early 1990s. In the time between its conception and release, nearly the entire lineup of the band had changed numerous times. It was once dubbed by *The*

New York Times "The Most Expensive Album Never Made" .*[63] Recorded in fourteen separate studios with reported production costs of $13 million, *Chinese Democracy* was eventually released in November 2008.*[64]

- ● *Detox*

Video games

- ● *Aliens: Colonial Marines*
  First announced in 2001, Aliens: Colonial Marines spent over 12 years in development hell. The original game which was announced in 2001 to be in development by Check Six Games, was cancelled and the rights for the Alien franchise were sold in 2006 to Sega.*[65] On December 15, 2006 Gearbox Studios announced they were developing Aliens Colonial Marines as a sequel to the 1986 film *Aliens*.*[66] The game spent another 7 years in development hell before it was released in 2013. Aliens: Colonial Marines has received mostly negative reviews. Most complaints in the negative reviews of the game included bugs, bad A.I., unbalanced gameplay, and low quality graphics in the single-player game as well as a crude and poorly implemented multiplayer cooperative mode. The game currently holds a Metacritic score of around 45%.*[67]*[68]

- ● *Duke Nukem Forever*
  The sequel to the 1996 game Duke Nukem 3D, Duke Nukem Forever, was in development hell for 14 years: from 1997*[69] to its release date in 2011. Changes of the game engine from the Quake II engine to the Unreal engine,*[69] conflict with Take-Two,*[70] and the bankruptcy of 3D Realms game studio*[71]*[72] caused the long development of the game. In 2010 Gearbox studios acquired the rights for the development*[73] and released Duke Nukem Forever in 2011.*[74]*[75] The game was critically disappointing upon release, with most of the criticism directed towards the game's clunky controls, long loading times, offensive humor, and overall aged and dated design. It holds a Metacritic score of around 50%.*[76]*[77]*[78]

- ● *Final Fantasy XV*
  Originally titled *Final Fantasy Versus XIII*, it was announced in 2006 as a spinoff of *Final Fantasy XIII* for PlayStation 3. Following a long period with little news on the game, it was re-announced as the next mainline installment of the series on PlayStation 4 and underwent large changes in direction such as making the game a self-contained story and replacing the main heroine.*[79]*[80] The game is now scheduled for a

worldwide release date in 2016, more than 10 years after it was initially announced.*[81]*[82]*[83]

- ● *Half-Life 2: Episode 3/Half-Life 3*
  After the release of Half-Life 2: Episode Two in 2007, concept art for the then presumed Half-Life 2: Episode 3 leaked online showing both characters and the lost Aperture Science ship *Borealis*.*[84] As of 2015, however, it is still unknown if a new Half-Life game is currently in development.

- ● *The Last Guardian*
  The Last Guardian was announced in 2007 to be in development at Team Ico.*[85] A short trailer released in 2007 shows a young boy who befriends a giant bird/cat-like creature. Creative conflicts between the developers and the publisher Sony, cause the game to remain in development hell, particularly after project lead Fumito Ueda left Sony but remained active in the game's development. Sony assured fans that the game was still in development over the next six years, but were sparse on further details, until June 2015 when the game was formally re-introduced as a PlayStation 4 title for release in 2016.

- ● *Mother 3*
  A sequel to the 1994 *Mother 2* (released as *Earth-Bound* in 1995 in North America). The game was initially intended to be released on the Super Famicom like its predecessor,*[86] before shifting focus to the Nintendo 64 Disk Drive. Following the failure of the Disk Drive, the game was shifted to the standard Nintendo 64,*[87] before the development team's inexperience with 3D-oriented video game creation and a large series of delays led to the game being quietly cancelled in 2000. Eventually, assets from the cancelled project were later collected and converted to a 2D format, and the project restarted development on the Game Boy Advance. Nine years after its conception, *Mother 3* was finally released on the Game Boy Advance in 2006, but only in Japan, though it received a well-regarded and highly successful fan translation to English.

- ● *Team Fortress 2*
  Was announced in 1999 and took 8 years to be released. With a complete change in gameplay and art direction, the North American release took place on 9 October 2007.*[88]

2.9.5 See also

- ● Turnaround (filmmaking)

- ● Vaporware

2.9.6 References

[1] Marx, Andy (February 28, 1994). "Interactive development: The new hell". *Variety* (New York) **354** (4): 1.

[2] Adler, Warren (October 3, 1999). "How My Novel Was Almost 'Developed' Into Oblivion". *New York Times*. p. AR11.

[3] "Cover Story: Writers Paid for Movies Never Made," Spillman, Susan. USA Today. McLean, Va.: January 16, 1991. pg. D1

[4] "Dept. of development hell," Kerrie Mitchell. Premiere. (American edition). New York: February 2005.Vol.18, Iss. 5; pg. 40

[5] "Books Into Movies: Part 2,"Warren, Patricia Nell. Lambda Book Report. Washington: April 2000.Vol.8, Iss. 9; pg. 9. (Best selling novel The Front Runner has spent over 25 years in development hell)

[6] McDonald, Paul & Wasko, Janet (2008) *Hollywood Film Industry*. Malden, MA: Blackwell Publishing. p. 54

[7] Lyons, Charles (2001) Development Hell freezing over? *Variety* 382(1). 1-71

[8] Schnepp, Jon (director) (2015). *The Death of "Superman Lives": What Happened?* (Documentary). Event occurs at 1:27:52.

[9] Paul W. S. Anderson, Lance Henriksen and Sanaa Lathan (2004). *Aliens vs. Predator. 20th Century Fox*.

[10] Kit, Borys (2012-01-05). "'Akira' Production Offices Shut Down As Warner Bros. Scrutinizes Budget (Exclusive)". *The Hollywood Reporter*.

[11] Britting, Jeff (2009). "Bringing *Atlas Shrugged* to Film". In Mayhew, Robert. *Essays on Ayn Rand's Atlas Shrugged*. Lanham, Maryland: Lexington Books. p. 195. ISBN 978-0-7391-2780-3.

[12] Bond, Paul (January 22, 2014). "'Atlas Shrugged: Who Is John Galt?' Starts Production With New Cast (Exclusive)". *The Hollywood Reporter*. Retrieved February 3, 2014.

[13] "Hollywood Curses". Xfinity. Retrieved November 7, 2013.

[14] "Mike Myers may return to "Austin Powers"". Entertainment Weekly. Retrieved 2015-04-26.

[15] "Mike Myers is Writing Austin Powers 4". Slashfilm. Retrieved 2015-04-26.

[16] "Toronto: Fleming Q&A's Mike Myers On 'Supermensch' Directorial Debut". Deadline. Retrieved 2013-09-16.

[17] Galbraith, Jane (1993-01-06). "'Costs Force Paramount to Delay Filming 'Beverly Hills Cop III'". *Los Angeles Times*. Retrieved 2010-09-29.

[18] Dutka, Elaine (1991-09-30). "Movies: Don Simpson and Jerry Bruckheimer just say no to Paramount's offer to make a third 'Beverly Hills Cop.'". *Los Angeles Times*. Retrieved 2010-09-29.

[19] Beck, Marilyn (1988-03-16). "Judge Reinhold Still Unsigned For `Beverly Hills Cop Iii`". *Chicago Tribune*. Retrieved 2010-09-29.

[20] "Beverly Hills Cop Iii' Could Finally Get Rolling This Summer". *Orlando Sentinel*. Retrieved 2010-09-29.

[21] David Hughes (March 2004). "The Dark Knight Strikes Out". *Tales From Development Hell*. London: Titan Books. pp. 192–211. ISBN 1-84023-691-4.

[22] Davidson, Paul (2003-07-23). "Sequel to *The Italian Job* Proposed". *IGN*. Retrieved 2008-10-08. Fleming, Michael and McNary, Dave (2004-07-19). "New man for the 'Job'". *Variety* (Reed Business Information). Retrieved 2008-09-17. McNary, Dave (2004-09-26). "Par reunites 'Job' crew". *Variety* (Reed Business Information). Retrieved 2008-09-17. Keck, William (2004-09-23). "'Huckabees' stars are all 'Heart'". *USA Today*. Retrieved 2008-10-10. See also: Davidson, Paul (2004-09-27). "New Italian Job Looks Likely". *IGN*. Retrieved 2008-09-17.

[23] McNary, Dave (2004-11-07). "Par: Déjà vu all over again". *Variety* (Reed Business Information). Retrieved 2008-10-05. See also: Tecson, Brandee (2005-12-27). "Mark Wahlberg Hits The Gridiron For Role In True–Life Tale 'Invincible'". *MTV*. Retrieved 2009-03-18.

[24] Davidson, Paul (2005-12-16). "'The Brazilian Job' Targets Summer 2007". *IGN*. Retrieved 2008-06-17. Davidson, Paul (2007-05-02). "*Brazilian Job* Still On". *IGN*. Retrieved 2008-06-17.

[25] Fleming, Michael (2005-05-31). "Par puts vet on the 'Job'". *Variety* (Reed Business Information). Retrieved 2008-09-11.

[26] Goldman, Eric (2007-08-17). "Exclusive Interview: Seth Green". *IGN*. Retrieved 2008-09-15.

[27] Corliss, Richard (September 12, 2013). "Dallas Buyers Club: McConaughey Shines as a Homophobe Who Gets AIDS". *Time*. Retrieved January 7, 2014.

[28] Whatever Happened to Foodfight? | Cartoon Brew

[29] "YouTube". YouTube. Retrieved 2013-09-08.

[30] "Foodfight!" Coming To DVD | Cartoon Brew

[31] Parfitt, Orlando (October 13, 2009). "Independence Day 2 News". IGN Entertainment. Retrieved October 13, 2009.

[32] "Fox's 'Independence Day 2' Moved From Busy 2015 Summer to 2016". FirstShowing.net. November 12, 2013. Retrieved 2013-11-12.

[33] "Paramount's Future- from 1985". YouTube. Retrieved 2014-06-05.

[34] Borys Kit (2007-05-09). "Future or past for Rodriguez?". *The Hollywood Reporter*. Archived from the original on May 25, 2007. Retrieved 2007-07-16.

[35] West, Scott (February 8, 2012). "Live-Action 'Jetsons' Movie Is Forward Again With New Writers". ScienceFiction.com. Retrieved December 30, 2014.

[36] Tony Fletcher's iJamming! . . .Keith Moon news, reviews and links

[37] http://www.alexcox.com/pdfs/MOON_1.pdf

[38] Sunday Mirror, 6, Aug. 2000

[39] The Lost Roles of Mike Myers | Splitsider

[40] Poor Scripts Hold Up Keith Moon Movie | FlicksNews.net

[41] AFM Briefs: 'The Hunted', 'April Apocalypse'; Alvernia & Fu Works in Poland

[42] Marlatt, Benjamin (July 19, 2015). "Who Knew Surf Rock Could Get So Complicated?". MoviePilot.

[43] DreamWorks Animation (December 10, 2010). "DreamWorks Animation Pioneers Groundbreaking Combination of CG and Hand-Drawn Animation Techniques in Me and My Shadow for March 2013". *DreamWorks Animation*. Retrieved November 2, 2012.

[44] DreamWorks Animation (March 8, 2011). "DreamWorks Animation Announces Feature Film Release Slate Through 2014". *DreamWorks Animation*. Retrieved November 2, 2012.

[45] McClintock, Pamela (June 11, 2012). "Stephen Colbert, Allison Janney Join Voice Cast of 'Mr. Peabody & Sherman' (Exclusive)". *The Hollywood Reporter*. Retrieved November 2, 2012.

[46] "DreamWorks Animation Pushes Back Release for 'Mr. Peabody & Sherman'". *The Hollywood Reporter*. February 5, 2013. Retrieved February 6, 2013.

[47] Brin, David (1998), *The Postman: An Impression by the Author of the Original Novel*, retrieved January 15, 2012

[48] Sin City 2 is a go... at last! | Flickering Myth

[49] David Hughes (2001). *The Greatest Sci-Fi Movies Never Made*. Independent Publishers Group. pp. 172–186. ISBN 1-55652-449-8.

[50] "Kevin Smith's *Superman Lives* cast". *Superman Homepage*. 1999-03-02. Retrieved 2008-02-04.

[51] Blizzard Entertainment (9 May 2006). "BLIZZARD ENTERTAINMENT® AND LEGENDARY PICTURES TO PRODUCE LIVE-ACTION WARCRAFT® MOVIE". *Web.Archive.org*. Archived from the original on 25 Nov 2007.

[52] Vejvoda, Jim (November 27, 2013). "Warcraft Movie Rescheduled to Avoid Star Wars: Episode VII". IGN.

[53] "Official X-Files 2 Announcement!". *IGN*. October 31, 2007. Retrieved September 5, 2009.

[54] Davidson, Paul (January 19, 2005). "Duchovny Hopes for a Couple More *X-Files*". *IGN*. Retrieved September 5, 2009.

[55] Webster, Christopher. "Exclusive review of Michael Bartlett's TIMELESS script!". *Quiet Earth*. Retrieved 13 October 2015.

[56] Hardawar, Devindra. "Concept Art For Time Travel Thriller Timeless". *Slashfilm*. Retrieved 13 October 2015.

[57] "Alexis Bledel to be in "Timeless"?". *Gilmore News*. Retrieved 13 October 2015.

[58] Webster, Christopher. "Michael Bartlett talks ZOMBIE DIARIES 2 and why you SHOULD NOT buy the US version". *Quiet Earth*.

[59] Stockman, Tom. "SLIFF 2014 Interview: Michael Bartlett – Director of TREEHOUSE". *wearemoviegeeks.com*.

[60] Ford, Rebecca. "Boundless Pictures Options 'Timeless' Script From 'Zombie Diaries' Writer (Exclusive)". *The Hollywood Reporter*. Retrieved 13 October 2015.

[61] "Filmography Timeless". *Boundless Pictures*. Retrieved 13 October 2015.

[62] Wolk, Douglas (October 31, 2011). "The Smile Sessions: The Story Behind the Box".

[63] Leeds, Jeff (March 6, 2005). "The Most Expensive Album Never Made". New York Times. Retrieved November 17, 2007.

[64] Powers, Ann. Review: *Chinese Democracy*. *Los Angeles Times*. Retrieved on April 9, 2010.

[65] http://www.ign.com/articles/2006/12/11/sega-hunts-down-alien

[66] *Game Informer*, March 2008, Issue 79, p. 49

[67] http://www.metacritic.com/game/xbox-360/aliens-colonial-marines

[68] http://www.metacritic.com/game/playstation-3/aliens-colonial-marines

[69] Wernicke, Brad (June 16, 1998). "George Broussard (06/16/98); on the switch from Quake II to Unreal engine for Duke Nukem Forever". *IGN*. Planet Duke. Archived from the original on December 16, 2005.

[70] Morris, Chris (June 11, 2003). "Duke Nukem vs. Take Two". *CNN*. Retrieved January 17, 2010.

[71] "Technology | Duke Nukem developer goes bust". BBC News. May 7, 2009. Retrieved July 21, 2009.

[72] Totilo, Stephen (May 18, 2009). "3D Realms: We're Not Closing, Spent $20 Million On Duke Nukem Forever". Kotaku. Retrieved May 18, 2009.

[73] Lee, Garnett (10 September 2010). "Talking Duke Nukem Forever With Gearbox Software's Steve Gibson". *Shack News* (Los Angeles CA). Retrieved 11 January 2013.: "Allen Blum and those guys, they're actually now in the Gearbox Software building on the tenth floor. We brought them in; they're now connected to the Gearbox infrastructure and our central team of animators and modelers and sound engineers."

[74] Cullen, Johnny (24 May 2011). "Hell freezes, pigs fly: Duke Nukem Forever goes gold". *VG24/7*. Retrieved 11 January 2013.

[75] Ewalt, David M. (February 11, 2011). "Duke Nukem Forever Balls of Steel Edition". *Forbes*.

[76] http://www.metacritic.com/game/pc/duke-nukem-forever

[77] http://www.metacritic.com/game/playstation-3/duke-nukem-forever

[78] http://www.metacritic.com/game/xbox-360/duke-nukem-forever

[79] "Final Fantasy XV: Stella is gone, Episode Duscae 2.0 slated for June 9 - Gematsu". Retrieved 2015-06-14.

[80] Sinha, Ravi. "Final Fantasy 15: Why Development Hell Can be The Quietest". *Gamingbolt*. Retrieved 6 January 2015.

[81] Brown, Peter (2015-08-06). "Final Fantasy 15 Release Date Confirmed for 2016". GameSpot. Archived from the original on 2015-08-06. Retrieved 2015-08-06.

[82] Woek, Kristofer (2015-08-06). "New Final Fantasy XV trailer released for Gamescom, 2016 release confirmed by director". Digital Trends. Retrieved 2015-08-06.

[83] Fahmy, Albaara (2015-08-11). "Final Fantasy 15 is aiming to release everywhere in the world at the same time". Digital Spy. Retrieved 2015-08-11.

[84] http://gamerant.com/halflife-2-episode-3-concept-art-dyce-160194

[85] http://multiplayerblog.mtv.com/2009/03/25/new-game-from-team-ico-will-be-like-ico-gdc-2009/

[86] http://www.1101.com/nintendo/nin13/nin13_2.htm

[87] http://www.1up.com/do/feature?pager.offset=1&cId=3154276

[88] Team Fortress 2(1998-2007): A Long Development Cycle Because of "Valve Time".

2.10 Duke Nukem 3D: Reloaded

Duke Nukem 3D: Reloaded (formerly known as *Duke Nukem Next-Gen*)[1] was a first-person shooter fan project in development. The title was announced on the Gearbox forums on October 13, 2010[3] and is based on the *Duke Nukem* series. The game was intended to be a next generation reimagining of the 1996 game *Duke Nukem 3D*.[4][5][6][7]

2.10.1 Development

In the fall of 2010, Frederik Schreiber had started throwing around the idea of doing a *Duke Nukem 3D* remake. Schreiber then created a test map to give an idea of what it may look like, which he then took screenshots of and posted on the Gearbox forums.[8] Shortly after posting the screenshots the images and the project made their way to various gaming sites.[9][10][11] causing a small buzz within the gaming community.

Schreiber went about getting the official permission to continue the project. He first contacted Gearbox Software, who told him to contact George Broussard and Scott Miller at 3D Realms. Schreiber proceeded to contact 3D Realms. The screenshots for the project were enough to convince Scott Miller to a certain degree about the project, but the game would need Take Two's permission for it to happen.

Schreiber again contacted Gearbox, hoping they would have a better relationship with Take Two than 3D Realms. After following the proper channels within Gearbox, he was able to get in contact with PJ Putnam, Vice President and General Counsel of Gearbox Software. Gearbox was interested in helping the project and Schreiber was eventually granted a "personal non-commercial license" to Duke Nukem.[12]

Having received official permission to proceed, Schreiber officially announced the game on October 13, 2010, under the name *Duke Nukem Next-Gen*, revealing he had set up a small team to work with. It was also stated the game would be based on the Unreal Engine 3 and would not require any other game for it to run.

On November 4, 2010, the game was officially renamed to *Duke Nukem 3D: Reloaded*.[1]

2.10.2 Release delayed

The game has been put on an indefinite hold as of September 24, 2011, pending the resolution of differences between the Interceptor Entertainment team and Gearbox Software due to ambiguity on whether or not the finished product would actually be allowed to see release.[13]

Schreiber spoke out on the delay in 2013, stating in an interview published in the July issue of the Danish video-game magazine, gameplay, that it was a direct result of *Duke Nukem Forever*'s disappointing reception. He elaborated,

"The problem was that *Reloaded*, in its then present state, was both a prettier and better game than *Forever* was. So they [Gearbox] could under no circumstances allow us [Interceptor] to publish it, show it, or do anything at all with it, because it would destroy the sales-opportunities they had left in *Forever*." *[14]

2.10.3   See also

- *Black Mesa* (video game) – a fan remake of *Half-Life*.

- *Rise of the Triad* (2013 video game) – another project by the same company, an official reboot of *Rise of the Triad: Dark War*.

- *Shadow Warrior* (2013 video game) - a reboot of the game *Shadow Warrior*.

2.10.4   References

[1] http://www.dukenukemreloaded.com/forum/viewtopic.php?f=9&t=116

[2] "Team". *Duke Nukem Reloaded web-site*. Retrieved 2011-04-21.

[3] http://gbxforums.gearboxsoftware.com/showthread.php?t=112564[]

[4] AFP (2010-10-05). "Fan-made Duke Nukem: Next Gen gets OK". *news.com.au*. Retrieved 2012-01-09.

[5] Devore, Jordan (2010-10-14). "Duke Nukem: Next-Gen fan project is a go". *Destructoid*. Retrieved 2012-01-09.

[6] Ransom-Wiley, James (2010-10-14). "Fan-made Duke Nukem 3D remake green-lighted by Gearbox". *Joystiq*. Retrieved 2012-01-09.

[7] Yin-Poole, Wesley (2010-10-14). "Fan-made Duke Nukem gets the go-ahead". *eurogamer.net*. Retrieved 2012-01-09.

[8] Duke Nukem 3D: Reloaded – The Gearbox Software Forums

[9] Michael McWhertor (2010-09-27). "We'll Play Duke Nukem Forever Before This Duke 3D Remake Is Done". *Kotaku*. Retrieved 2012-01-27.

[10] Zimmerman, Conrad (2010-09-27). "Duke Nukem 3D fan-made remake screens impress". *Destructoid*. Retrieved 2012-01-09.

[11] Ridgeley, Sean (2010-09-27). "Duke Nukem 3D fan remake looks sick". *neoseeker*. Retrieved 2012-01-09.

[12] The Gearbox Software Forums

[13] Schreiber, Frederik. "Public Announcement". *Project Lead*. Interceptor Entertainment. Retrieved 24 September 2011.

[14] Jesper Krogh Kristiansen (20 June 2013). "Udviklerprofil: Interceptor Entertainment". *gameplay*.

2.10.5   External links

- Official website

- Official announcement on the Gearbox forums

- Kotaku story about the game's official announcement

2.11   Duke Nukem Forever

Duke Nukem Forever is a first-person shooter video game for Microsoft Windows, OS X, PlayStation 3, and Xbox 360. It is a sequel to the 1996 game *Duke Nukem 3D* as part of the long-running *Duke Nukem* video game series. It started development at 3D Realms and Triptych Games, and was finished by Gearbox Software and Piranha Games.

The game stars the titular action hero who must come out of retirement and save the world from aliens when they begin kidnapping the women of Earth. Intended to be groundbreaking, *Duke Nukem Forever* became a notable example of vaporware due to its severely protracted development schedule; the game was released in 2011 after fifteen years of development. Reception to *Duke Nukem Forever* was generally negative, with many critics singling out the game's simplistic mechanics, unpolished performance, second-rate graphics and dated humor.

2.11.1   Gameplay

Duke Nukem Forever is an action-oriented first-person shooter. Players takes control of Duke Nukem and navigate a series of levels which take place on Earth and beyond. The game allows players to interact with various in-game objects, including urinals and water coolers, as well as whiteboards, which allow players to draw their own images. The gameplay is similar in some respects to the first-person shooter games of the late 1990s and early 2000s, with each level culminating in a boss battle in which Duke has to fight and kill a large, significant alien.

Unlike in the previous games, Duke can only hold two weapons at any one time in a manner similar to the *Halo series*, although pipe bombs and laser tripwires are considered inventory items and as such are not limited by this restriction. The PC version allows Duke to hold four weapons in the single-player campaign. Items that have an effect on Duke can be picked up by the player; these items are steroids, beer, and the holoduke. Steroids increase the strength of Duke's melee attacks by a great deal for a limited time. Beer makes Duke much more resistant to damage, but blurs the screen. The holoduke creates a hologram of Duke Nukem that looks and acts in a very similar way to Duke, but often says slightly twisted versions of his one-liners. Whilst the holoduke is in effect, Duke becomes invisible and the AI characters do not recognize his presence. The jetpack also returns, but only in multiplayer.

Instead of the health system featured in the previous *Duke Nukem* games, in which health would be depleted when Duke was injured and would only increase upon finding a health pack, drinking water from fountains/broken fire hydrants, urinating, or using the portable medkit item, *Duke Nukem Forever* employs a system involving an "ego bar". The ego bar depletes when Duke is attacked; once it is fully depleted, Duke becomes susceptible to damage. Dying will result upon taking too much damage and cause the game to reload the last checkpoint. If the player avoids further damage, then the ego bar is restored as is Duke's health. The player can increase the size of the ego bar (thus increasing the amount of damage Duke can take) by interacting with certain objects throughout the game (for example, a mirror), and by defeating bosses. The game employs a save system that is solely based on checkpoints.

2.11.2 Plot

Twelve years after he saved the Earth from an alien invasion, Duke Nukem is a worldwide icon, and has achieved great fame from his heroic deeds. After sampling a video game based on his past heroics (the game Duke plays is a revamped version of the final level of the third episode of *Duke Nukem 3D*), he arrives on the set of a talk show for an interview. On his way to the show, Duke witnesses a news broadcast announcing that aliens have once again invaded. Unlike previous encounters, the aliens initially appear peaceful and at first seem to pose no harm to the humans of Earth.

Duke's talk show appearance is cancelled to allow television stations to cover the alien invasion, and he retires to the "Duke Cave", his personal home. There, Duke receives a call from the President and General Graves of the Earth Defense Force (EDF). The President orders him not to harm the invaders, and adds that he is in diplomatic talks with the alien overlord. Duke obliges this request, but remains uneasy about the whole situation from start. Before he can leave his chambers, he is attacked by hostile aliens who are swearing revenge on Duke.

Duke is forced to disobey the president's orders and fight his way through the alien hordes in an effort to save Earth. Whilst fighting through his casino, Duke witnesses the aliens abducting women, including his two live-in pop star girlfriends. Graves tells Duke that the women are being held

in the Duke Dome, and that the aliens have a vendetta to settle with Duke. He also warns Duke that the aliens are using the Hoover Dam to power a wormhole so more aliens can come through. Duke travels to the Duke Dome, using a wrecking ball to damage the building to gain access. Inside, he finds swarms of Octabrains and the missing women, who have been impregnated with alien spawn; Duke's girlfriends die after giving "birth" to alien babies. Duke finds the Alien Queen in control of the Duke Dome and kills her, but is wounded in the process and blacks out.

After regaining consciousness, Duke fights Pigcops and aliens in through the Duke Burger. Soon, he travels to the Hoover Dam in his monster truck; after battling through the dam, he finds his old friend Dylan, mortally wounded. He tells Duke that the reborn Cycloid Emperor is at the dam, and that the only way to shut down the portal is to completely destroy the dam. Before dying, he gives Duke his demolition charges and wishes him luck. Duke places the explosives and destroys the dam, but the currents nearly drown him.

Duke is revived by an EDF soldier, and awakens to find the portal gone. The President, who was also at the dam, rages at Duke for ruining his plans to work with the Cycloid Emperor, revealing that the President was actually intending to have the aliens kill Duke and he would cooperate with Cycloid Emperor so he could control the Earth, and that he has ordered a nuclear strike at the site of the dam to wipe out the remaining aliens, intending to leave Duke there to die as revenge for foiling his plans. The Cycloid Emperor emerges and kills the President and his security detail, revealing that he intended to kill the President after the deal. Duke kills the Cycloid Emperor and is rescued by Graves just as the nuclear bomb explodes.

The game ends with a satellite surveying the detonation area and listing Duke Nukem as killed in action, to which Duke replies off-screen, "What kind of shit ending is that? I ain't dead. I'm coming back for more!" In a post-credits scene, a short video depicts a press conference, where Duke announces his intent to run for the 69th President of the United States.

The Doctor Who Cloned Me

In the downloadable content *The Doctor Who Cloned Me*, Duke wakes up after the nuclear explosion and finds himself alive but trapped in a strange laboratory while video recordings of himself declaring his bid for Presidency play on monitors. After escaping, Duke discovers that not only are the aliens continuing their invasion, but his old nemesis Dr. Proton (the villain of the original *Duke Nukem* computer game) has returned and is building an army of robotic Duke clones to fight the aliens and conquer Earth himself.

Duke infiltrates Proton's laboratory in Area 51 by posing as one of the clones. Eventually, Proton spots him and attacks Duke but he escapes and is reunited with Dylan (revealed as still alive). With Dylan's help, Duke locates and kills Dr. Proton. General Graves then communicates with Duke to inform him that the aliens are being bred by an Alien Empress that is nesting on the moon. After finding a teleporter leading up to the moon, Duke commandeers a moon rover and destroy the Alien Empress, saving Earth and its women once again.

2.11.3 Development

Main article: Development of Duke Nukem Forever

Originally in development under 3D Realms, director George Broussard, one of the creators of the original *Duke Nukem* game, first announced the title's development in April 1997, and various promotional information for the game was released between 1997 and 1998. After repeatedly announcing and deferring release dates, 3D Realms announced in 2001 that it would be released simply "when it's done" . No official video of the game was shown for almost 8 years, until 3D Realms released a new teaser trailer in December 2007, but the game "sank" yet again soon afterwards.

In May 2009, 3D Realms was downsized for financial reasons, resulting in the loss of the game's development team. Statements by the company indicated that the project was due to go gold soon with pictures of final development. Take-Two Interactive, which owns the publishing rights to the game, filed a lawsuit in 2009 against 3D Realms over their failure to finish development. 3D Realms retorted that Take-Two's legal interest in the game was limited to their publishing right. The case was settled with prejudice and details undisclosed in July 2010.

On September 3, 2010, after 14 years, *Duke Nukem Forever* was officially reported by 2K Games to be in development at Gearbox Software. It was originally confirmed to be released on May 3, 2011 in North America, with a worldwide release following on May 6, 2011.[8] This was, however, delayed by a month to June 10 internationally, with a North American release on June 14.

On May 24, 2011, it was announced that *Duke Nukem Forever* finally "went gold" after 15 years.[9][10] After going gold the launch trailer for *Duke Nukem Forever* was released on June 2, 2011, quelling any doubt that release was anything but imminent.[11] On June 27, 2011, Aspyr Media announced that *Duke Nukem Forever* would be making its way onto Mac OS X in August 2011. It was made available for pre-order on June 27 via their online game distribution platform GameAgent.[12]

2.11.4 Marketing

Promotion at the E3 2011

Duke's First Access Club is joined by using a code, obtained from either the pre-order of the game, the *Borderlands Game of the Year Edition*, or *Borderlands* on Steam (if bought before the club was announced), on the *Duke Nukem Forever* website. Members are granted access to the demo, wallpapers, concept art, artwork, podcasts (which are added often), the theme song, and screenshots.

Emails were sent asking members to "please help [Gearbox] obtain the most accurate up to date information for your First Access profile." Members were then prompted to choose their preferred platform of choice for the *Duke Nukem Forever* demo by May 15, 2011. It then stated that "users that currently live in a territory where the demo may not be supported on console will automatically be defaulted to the PC Steam option." [13] Gearbox sent a second e-mail to First Access Members in conjunction with a video showing that the *Duke Nukem Forever* demo was released on June 3, 2011.[14]

A special limited Collector's Edition was available upon release called "Balls of Steel Edition" for all platforms. This version includes a five-inch bust of Duke Nukem, a 100-page hardcover artbook following the development of the game, postcards, sticker, a comic book, playing cards, dice, poker chips and foldable papercraft, and with every item being marked with the *Duke Nukem Forever* logo.[15][16] Another edition called the "King Edition" was made available exclusively for pre-order from EB Games in Australia and New Zealand. It comes with the bonus "Ego Boost" , Duke Playing Cards and Duke Bubblegum.[17]

Two themes, avatar items, and a gamerpic pack are available for download for the Xbox 360 and PlayStation 3. The premium theme for the Xbox 360 showcases the inside of Duke Burger during the alien invasion. The avatar items for said system include Duke's outfit, his throne, the Freeze Ray, a Pigcop mask, and a pet Octabrain, whilst the gamerpic pack features "babes, aliens, and the King himself." The official *Duke Nukem Forever* website hosts the free PlayStation 3 theme, which includes three wallpapers and an icon set.*[18]

2K Games launched a website titled "Boob Tube" to promote the game. The website features videos and features to download. On May 19, 2011, a flash game was released on the website titled *Duke Nudem* where players have to shoot targets against a CPU bot "woman" of their choice, and if successful will have a piece of clothing taken off the girl until she is topless. However, if the player loses, the actress will act as though Duke has stripped naked.*[19] Additionally 2K released for iOS a *Duke Nukem Forever Soundboard* which includes a number of Duke Nukem's phrases to be played back.*[20]

Originally set for release in Australia on June 10, 2011, the game was made available for sale a day early on June 9 from all retailers due to street date being broken.*[21]

2.11.5 Downloadable content

Duke's Big Package

In North America, video game retailer GameStop promised exclusive in-game content for customers pre-ordering *Duke Nukem Forever*. The exclusive content, known as "Duke's Big Package" allowed the player from the start of the game to access "Big Heads", the "Ego Boost", and custom ingame T-shirts. A code printed on the final receipt could, at the time of release, be activated over Xbox Live, PlayStation Network, and Steam.*[22]*[23]

Hail to the Icons Parody Pack

Duke Nukem Forever: Hail to the Icons Parody Pack, contains three new game modes, and four new multiplayer maps, each with new weapons. It is available on Xbox Live, Steam, and PlayStation Network. It was released on October 11, 2011.*[24]

The Doctor Who Cloned Me

Duke Nukem Forever: The Doctor Who Cloned Me includes an all new single-player campaign which features the return of Duke's nemesis from the original *Duke Nukem*, Dr. Proton.*[25] It includes new weapons, enemies, and bosses. The DLC also includes four new multiplayer maps. The DLC was released on December 13, 2011.*[26] It holds a score of 52/100 on Metacritic for PC*[27] and 58/100 for

Xbox 360.*[28] Gamespy rated it a 1.5/5,*[29] OXM rated it 4/10*[30] and Eurogamer rated it 5/10 and stated "Duke's trying his best, but there's still too much of the past hanging around and holding him back." *[31]

2.11.6 Reception

Critical reception

Duke Nukem Forever was critically disappointing upon release, with most of the criticism directed towards the game's clunky controls, long loading times, offensive humor, and overall aging and dated design. Aggregating review websites GameRankings and Metacritic calculated the Xbox 360 version 49.36% and 49/100,*[32]*[37] the PlayStation 3 version to be 47.6% and 51/100*[34]*[36] and the PC version 48.52% and 54/100.*[33]*[35] Elton Jones of *Complex* chose the game as one of "the most disappointing games of 2011".*[55] Jim Sterling, review editor for Destructoid, said that this game was "like a disease" *[39] and named it the "shittiest game of 2011".*[56] Ben "Yahtzee" Croshaw, creator of Zero Punctuation, listed it as #2 on his list of the worst games of 2011, losing to both *Battlefield 3* and *Call of Duty: Modern Warfare 3*.*[57]

Many critics took issue with the level design and shooting mechanics, particularly when compared to both the original *Duke Nukem 3D* and those of other modern-day shooters. Kevin VanOrd of GameSpot felt that the "joy of that game's shooting has been flattened" with "little sense of impact", finding the overall design to be "tedious", and ended his video review by calling Duke Nukem Forever a "bad, boring, bargain bin kind of game".*[45] Eurogamer commented that "few of the locations [inspired] the sort of exploration and excitement that made *Duke 3D* such a memorable experience. *Duke Nukem Forever* is linear to a fault, and huge chunks of the game are spent simply walking from one fight to another through uninspired corridors." *[41] IGN criticized "the frequent first-person platforming segments that make up an unnecessarily large percentage of the story mode", although they stated the "shooting sections are simple fun".*[50] GamesRadar concluded that "*Duke Nukem Forever* 's world-record development time has produced an ugly, buggy shooter that veers back and forth between enjoyably average and outright boring, with occasional surges of greatness along the way." *[48] GamePro felt that "Unexpected moments ... are really the game's biggest strengths. But they're few and far between." *[43] X-Play gave the game a 1 out of 5, criticizing the graphics, load time, number of enemies onscreen, the multiplayer, being called "an afterthought", the game's "creepy, hateful view of women." ,*[54] and the hive level, with Adam Sessler saying that "this is all played for laughs" .

Many reviewers questioned the design choices in comparison to *Duke Nukem 3D*, with Kotaku stating that "Old-school shooters, and this is definitely trying to be one of those with its basic AI and lack of cover mechanics, always had two great things going for them: speed and a ridiculous arsenal of weapons... Forever eschews this in favour of a plodding pace and two guns." *[58] Noting its negative mix with modern shooter conventions, The Escapist agreed: "having been almost cryo-frozen for more than a decade, then awoken and peppered with modern touches, [Duke Nukem Forever] feels so out of place." *[59]

Another common criticism was with the game's lack of technical sophistication, including inconsistent graphics and unacceptably long loading times, which GameTrailers called "unholy";*[49] Eric Neigher of GameSpy found the console versions took up to 40 seconds to load a level.*[47] He also criticized the game's multiplayer mode for running so slowly, no one can play it without experiencing large lag spikes.*[47] *Edge* commented that "the myriad technical shortcomings – particularly prevalent on the console ports – only get worse the further you progress into the campaign" ,*[40] a view echoed by Game Revolution: "when they started on the design, that tech was already outdated" .*[44] The PC version has since been patched to greatly decrease loading times and to add two optional inventory slots.

The use of the series' trademark humor received a mixed response. In one regard, some critics such as Team Xbox praised the voice work of Jon St. John, who did an "excellent job as always with Duke's persona" ,*[60] whilst others such as Machinima.com*[61] appreciated the comedic gameplay tips and pop culture references; however, the same critic also noted that "parts of the narrative and dialogue show clear evidence of the game's elongated development. Many pop culture references refer to media in the early 2000s, with one-liners co-opted from 'guy' movies like *Old School*, *Highlander*, and *Commando*, which in itself could cause blank stares from most of the current potential audience."*[61] Australian gaming website PALGN felt the game was "saved only by its humor and nostalgic value." *[52] *Official Xbox Magazine* UK thought that the humor "isn't so much offensive or misogynistic as just suffering from an adolescent fixation with boobs and crowbarred-in innuendo." Joystiq noted that the game's multiplayer mode "Capture the babe" capture the flag, involving "spanking a woman into submission" , "really is as painful as it sounds" .*[51] Many critics conversely criticized the characterization of Duke Nukem, declaring his decidedly one-dimensional personality juvenile and outdated in comparison to more recent video game heroes.

One particular section that received considerable criticism is the hive level, in which Duke encounters abducted women who have been forcibly impregnated with aliens. Duke has to kill them before the alien's birth does so. Both the level itself and the inclusion of disembodied, slappable "wall boobs" were listed in GamesRadar's "8 worst moments in *Duke Nukem Forever*".*[62] *OXM* noted that it "doesn't mesh with the rest of the game's tone at all", and the fact that Duke remains unfazed and continues to crack jokes about the situation was considered "outright revolting", which led to labelling Duke a "thoroughly detestable psychopath" by 1UP and Destructoid respectively.*[38]*[39] Zero Punctuation noted that the level is "as jarring a shift of tone as you can get without splicing five minutes of *The Human Centipede* into the middle of *Mallrats*." *[63]

Quite a few critics cited the long and fragmented development time as a major factor in the finished product. In a positive review *PC Gamer* noted that "years of anticipation will spoil *Duke Nukem Forever* for some" , adding, "There's no reinvention of the genre here, no real attempt at grandeur... Check unrealistic expectations at the door and forget the ancient, hyperbolic promises of self-deluded developers" , and concluded, "Don't expect a miracle. Duke is still the hero we love, but struggles to keep up with modern times." *[53] Game Informer, whilst disappointed in the game concluded "I'm glad Gearbox stepped up and finished this game, but after hearing about it for 12 years, I have no desire to relive any of it again. I'm now satisfied in my knowledge of what Duke Nukem Forever is and ready to never talk about it again. Welcome back, Duke. I hope your next game (which is teased after the credits) goes off without a hitch." *[42] GiantBomb however concluded that for those "part of that faction that finds yourself so fascinated by this whole project that you need to know how it ends, I recommend you play *Duke Nukem Forever* for yourself. But I'd practically insist that you do so on the PC and try to wait for a sale. If you're not willing to play a sloppy, cobbled together first-person shooter just because it has some kind of weird historical meaning, though, just forget this ever happened and move on." *[64] Jake Denton of *Computer and Video Games* wrote that parts of the game were fun to play and listed it as one of the "5 most underrated games of 2011" , while admitting the game's overall faulty structure.*[65] Also Joseph Milne of *FPSguru.com* featured the game on his list of "Top 5 underrated games" at number 4 on the list.*[66]

Sales

According to research firm NPD, *Duke Nukem Forever* sold 376,300 units in its first month (sales results do not include digital copies).*[67] Take-Two Interactive, the parent company of 2K Games, revealed in July 2011 that the game sales were half of their initial expectations.*[68] However, in an earnings call on August 8, 2011, Take-Two stated that *Duke Nukem Forever* would prove to be profitable for the company.*[69]

2.11.7 References

[1] Broussard, George; Blum, Allen H., III (September 3, 2010). "Duke Nukem Forever Hands-on Preview (comment from George Broussard)". Retrieved February 11, 2011. Triptych Games which continued the game for us through all of 2009 and into 2010 with Gearbox. Triptych is made up of 9 3DR employees who refused to let the game go and we found a way through the legal maze to keep them working on the game and to keep the game alive. They have been the development force for the last year that's made the game possible. What you see coming from PAX right now is what we originally made at 3DR with polish and additional work by Triptych and assistance from Gearbox

[2] Hackersho, Yu Yu (September 3, 2010). "Duke Nukem Forever hitting in 2011". Gameinformer.com. Retrieved February 18, 2011. We now know that Gearbox started working on the game year ago

[3] "Duke Nukem Forever Interview with Gearbox Software". AusGamers.com. Retrieved February 18, 2011.

[4] "Press Release: Duke Nukem Forever Set to Kick Ass and Chew Bubblegum". DukeNukem.com. January 21, 2011. Retrieved February 18, 2011. 2K Games and Gearbox Software announced today what will be a landmark date in gaming history...

[5] "3D Realms Forum - Features of the DNF engine?". June 14, 2008. Retrieved June 14, 2012. Unreal. I believe we branched off somewhere around the Unreal 2 time when they added static meshes. Since then we've redone the rendering 100% and it's a fully modern engine.

[6] Robert Purchese. "Gearbox delays Duke Nukem Forever". Eurogamer.

[7] DNF japan

[8] Jeff Cork (January 21, 2011). "Exclusive: Duke Nukem Forever Has A Release Date". Game Informer. Retrieved January 21, 2011.

[9] Andy Robinson (March 24, 2011). "Duke Nukem Forever delayed again (really)". Retrieved March 24, 2011.

[10] Gearbox Twitter (May 24, 2011). "Duke Nukem Forever gone gold". Retrieved May 24, 2011.

[11] DigitalTrends (June 2, 2011). "Against all odds and logic, Duke Nukem Forever has a launch trailer". Retrieved June 2, 2011.

[12] "Duke Nukem Forever announced for Mac OS X". Blog.gameagent.com. June 27, 2011. Retrieved July 13, 2011.

[13] "Duke Nukem's First Access Club E-mail Hints Upcoming Demo Arrival". Game Focus.

[14] Chris Pereira. "Duke Nukem Forever Demo Coming on June 3". 1up.com.

[15] JC Fletcher. "Duke Nukem Forever 'Balls of Steel' Edition: Is this some kind of bust?". Joystiq.

[16] Andy Chalk. "2K Announces Duke Nukem Forever Balls of Steel Edition". The Escapist.

[17] "Duke Nukem Forever: King Edition "exclusive" at EBGames – PS3 News | MMGN Australia". Ps3.mmgn.com. Retrieved May 12, 2011.

[18] Elizabeth Tobey. "Deck out your 360 and PS3 with Duke Nukem Forever Goodness". 2K Games.

[19] "Shoot Targets to Get Duke Nukem Girls Topless". Kotaku. May 18, 2011.

[20] Crecente, Brian (May 27, 2011). "Duke Nukem Had Eggs For Breakfast, Your Mom Had Sausage". Kotaku.

[21] "Duke Nukem Forever Breaks Street Date". Kotaku. June 9, 2011.

[22] "Dukes Big Package" (PDF). GameStop. June 14, 2011.

[23] "Code Redemption Instructions". GameStop. June 14, 2011.

[24] Mike Fahey (October 11, 2011). "Oh Good, the Duke Nukem Forever DLC is Here". Kotaku. Retrieved March 23, 2014.

[25] Rossignol, Jim. "Also: Dukem Nukem DLC On Tuesday". Rock, Paper, Shotgun. Retrieved December 14, 2011.

[26] Charles Onyett (December 9, 2011). "Duke Nukem Forever Single-Player DLC Incoming". IGN. Retrieved March 23, 2014.

[27] "Duke Nukem Forever: The Doctor Who Cloned Me PC on Metacritic". Metacritic. Retrieved March 23, 2014.

[28] "Duke Nukem Forever: The Doctor Who Cloned Me for Xbox 360 on Metacritic". Metacritic. Retrieved March 23, 2014.

[29] Dan Stapleton (December 18, 2011). "DNF: The Doctor Who Cloned Me Review". GameSpy. Retrieved March 23, 2014.

[30] Ryan Mccaffrey (January 3, 2012). "Duke Nukem Forever: The Doctor Who Cloned Me review". Official Xbox Magazine. Retrieved March 23, 2014.

[31] Christian Donlan (December 20, 2011). "Duke Nukem Forever: The Doctor Who Cloned Me Review". Eurogamer. Retrieved March 23, 2014.

[32] "Duke Nukem Forever (Xbox 360) reviews at". GameRankings. June 14, 2011. Retrieved June 14, 2011.

[33] "Duke Nukem Forever (PC) reviews at". GameRankings. June 14, 2011. Retrieved June 20, 2011.

[34] "Duke Nukem Forever (PlayStation 3) reviews at". GameRankings. June 20, 2011. Retrieved June 20, 2011.

[35] "Duke Nukem Forever (PC) reviews at" . Metacritic. June 14, 2011. Retrieved June 20, 2011.

[36] "Duke Nukem Forever (PlayStation 3) reviews at" . Metacritic. June 14, 2011. Retrieved June 20, 2011.

[37] "Duke Nukem Forever (Xbox 360) reviews at" . Metacritic. June 14, 2011. Retrieved June 20, 2011.

[38] "Duke Nukem Forever Review for PC, 360, PS3 from 1UP.com" . *1UP.com*. June 14, 2011. Archived from the original on 2013-03-19. Retrieved June 14, 2011.

[39] Jim Sterling (13 June 2011). "Review: *Duke Nukem Forever*". Destructoid. Retrieved 13 July 2011.

[40] "Duke Nukem Forever review – Edge Magazine" . Nextgen.biz. Retrieved July 13, 2011.

[41] "Duke Nukem Forever" . Eurogamer. June 11, 2011. Retrieved June 11, 2011.

[42] Reiner, Andrew (June 14, 2011). "Duke Nukem Forever review: 12 Years In The Making.." . Game Informer. Retrieved June 17, 2011.

[43] post a comment. "Duke Nukem Forever Review from" . GamePro. Archived from the original on December 1, 2011. Retrieved July 13, 2011.

[44] KevinS (21 June 2011). "This took 14 years? Seriously?". GameRevolution. Retrieved 19 May 2015.

[45] "Duke Nukem Forever (PC) reviews at" . GameSpot. June 14, 2011. Retrieved June 14, 2011.

[46] "Duke Nukem Forever (Xbox 360) reviews at" . GameSpot. June 14, 2011. Retrieved June 14, 2011.

[47] Eric Neigher (14 June 2011). "Say it ain't so, Duke. Say it ain't so." . GameSpy. Retrieved 19 May 2015.

[48] Reparez, Mikel (June 14, 2011). "Duke Nukem Forever review" . Future Publishing. Retrieved June 15, 2011.

[49] "Duke Nukem Forever Review HD" . *GameTrailers*. June 16, 2011. Retrieved June 16, 2011.

[50] "Duke Nukem Forever" . IGN. June 11, 2011. Retrieved June 11, 2011.

[51] Nelson, Randy (June 10, 2011). "Duke Nukem Forever review: Fail to the King, Baby" . Joystiq. Retrieved June 10, 2011.

[52] "Duke Nukem Forever" . PALGN. June 11, 2011. Retrieved June 11, 2011.

[53] "Duke Nukem Forever" . PC Gamer. June 11, 2011. Retrieved June 11, 2011.

[54] Jason D'Aprile (21 June 2011). "*Duke Nukem Forever* Review" . X-Play. Retrieved 20 May 2015.

[55] Jones, Elton (December 24, 2011). "The 10 Most Disappointing Games of 2011" . Complex. Retrieved March 15, 2012.

[56] Jimquisition: The Ten Worst Games of 2011. YouTube (2012-06-26). Retrieved on 2013-07-31.

[57] The Escapist : Video Galleries : Zero Punctuation : Top 5 of 2011. Escapistmagazine.com. Retrieved on 2013-07-31.

[58] "Duke Nukem Forever: The Kotaku Review" . Kotaku. June 21, 2011. Retrieved June 21, 2011.

[59] Duke Nukem Forever Review (June 13, 2011). "The Escapist : Duke Nukem Forever Review" . Escapistmagazine.com. Retrieved July 13, 2011.

[60] "Sparky" (17 June 2011). "*Duke Nukem Forever* Review (Xbox 360)". Team Xbox. Retrieved 30 May 2015.

[61] Rob Smith (20 June 2011). "*Duke Nukem Forever* Review" . Machinima.com. Retrieved 30 May 2015.

[62] Reparaz, Mikel (June 21, 2011). "The 8 worst moments in Duke Nukem Forever" . Retrieved September 7, 2011.

[63] Croshaw, Ben (June 22, 2011). "Zero Punctuation: Duke Nukem Forever (for real this time)" (Video). *The Escapist*. Retrieved September 6, 2011.

[64] "Review: Duke Nukem Forever" . GiantBomb. Retrieved December 12, 2011.

[65] Denton, Jake (December 16, 2011). "5 most underrated games of 2011: Hidden gems from the last year..." . ComputerAndVideoGames.com. Retrieved March 16, 2012.

[66] The List: Top 5 Underrated Games. Fpsguru.com (2011-08-11). Retrieved on 2013-07-31.

[67] "Duke Nukem Forever Sales Results" . *IGN*. July 14, 2011. Retrieved September 7, 2011.

[68] "Take Two Estimates Lowered After Disappointing Duke Sales" . Gamasutra. July 5, 2011. Retrieved July 10, 2011.

[69] "Duke Nukem profitable, L.A. Noire ships 4 million says Take-Two" . PlayStation Universe. August 9, 2011. Retrieved August 10, 2011.

2.11.8 External links

- Official website
- *Duke Nukem Forever* news archive at 3D Realms web site
- *Duke Nukem Forever* at the Internet Movie Database

2.12 Fluorescent multilayer card

The **Fluorescent Multilayer Card** (**FMC**) is a hypothetical memory card technology that applies the same 3D optical data storage mechanism as the Fluorescent Multilayer Disc. The concept was developed by Constellation 3D.

An FMC has a square, transparent window with multiple layers beneath it. Each layer is made of a different fluorescent material that emits light; each layer emits light of a different color (wavelength). An FMC reader could distinguish one layer from another by the wavelength, and therefore read from them without the problems that limit the number of layers in a DVD.

Constellation 3D estimated that a card with a data area of 25 millimetres (0.98 in) could hold 10 gigabytes of data.

2.13 Fluorescent Multilayer Disc

Fluorescent Multilayer Disc (**FMD**) is an optical disc format developed by Constellation 3D that uses fluorescent, rather than reflective materials to store data. Reflective disc formats (such as Compact Disc and DVD) have a practical limitation of about two layers, primarily due to interference, scatter, and inter-layer cross talk. However, the use of fluorescence allowed FMDs to operate according to the principles of 3D optical data storage and have up to 100 data layers. These extra layers potentially allowed FMDs to have capacities of up to a terabyte, while maintaining the same physical size of traditional optical discs.

2.13.1 Operating principles

The pits in an FMD are filled with fluorescent material. When coherent light from the laser strikes a pit the material glows, giving off incoherent light of a different wavelength. Since FMDs are clear, this light is able to travel through many layers unimpeded. The clear discs, combined with the ability to filter out laser light (based on wavelength and coherence), yield a much greater signal-to-noise ratio than reflective media. This is what allows FMDs to have many layers. The main limitation on the number of layers in an FMD is the overall thickness of the disc.

2.13.2 Development

A 50 GB prototype disc was demonstrated at the COMDEX industry show in November 2000. First generation FMDs were to use 650 nm red lasers, yielding roughly 140 GB per disc. Second and third generation FMDs were to use 405 nm blue lasers, giving capacities of up to a terabyte.

D Data, Inc. acquired the patent portfolio of Constellation 3D in 2003, with plans to reintroduce the technology under the new name Digital Multilayer Disk (DMD).

2.13.3 See also

- Fluorescent Multilayer Card

- Hyper CD-ROM

2.14 Glaze3D

Glaze3D was a family of graphics cards announced by BitBoys Oy on August 2, 1999 that would have produced substantially better performance than other consumer*[1] products available at the time. The family, which would have come in the **Glaze3D 1200**, **Glaze3D 2400** and **Glaze3D 4800** models, was supposed to offer full support for DirectX 7, OpenGL 1.2, AGP 4×, 4× anisotropic filtering, full-screen anti-aliasing and a host of other technologies not commonly seen at the time. The 1.5 million gate*[1] GPU would have been fabricated by Infineon on a 0.2 μm eDRAM process,*[1] later to be reduced to 0.17 μm with a minimum of 9 MB of embedded DRAM and 128 to 512 MB of external SDRAM. The maximum supported video resolution was 2048×1536 pixels.

2.14.1 Development history

The Glaze3D family of cards were developed in several generations, beginning with the original Glaze3D "400" with multi-channel RDRAM instead of internal eDRAM. This was offered only as IP but with no takers. Bitboys revised the design and decided to have it manufactured themselves, in cooperation with Infineon Technologies, the chip fabrication arm of Siemens. They came up with a new **Glaze3D** pitched for release in Q1, 2000. The card promised extremely high performance compared to contemporary consumer GPUs. As bug-hunting, validation and manufacturing problems delayed the launch, new features became necessary and a DX7 variant with built-in hardware Transform & Lighting was announced, but never appeared.

The GPU was later redesigned under a new codename, **Axe**, to take advantage of DirectX 8 and compete with a developing competition. The new version sported such features as an additional 3 MB of eDRAM, proprietary Matrix Antialiasing and a vastly improved fillrate, as well as offering a programmable vertex shader and widened internal memory bus. The new card was to have been released as **Avalanche3D** by the end of 2001.

The third development, codenamed **Hammer**, started development as Axe lost viability toward the end of 2001. This new card was to be a high-end DirectX 9 part, offering new features such as occlusion culling, improved rendering performance and various other innovations. This version, like the ones before it, never shipped commercially.

Bitboys turned to mobile graphics and developed an accelerator licensed and probably used by at least one flat panel display manufacture, although it was intended and designed primarily for higher-end handhelds. Later on ATI bought Bitboys for an extra research and development unit, so as of 2008 Bitboys was owned by AMD. In 2009, Bitboys was transferred to Qualcomm.

2.14.2   Performance claims

A publicity screenshot designed to highlight the realism that Glaze3D cards were supposed to achieve.

The Glaze3D family was well known for the bold performance claims that were associated with it. The low-end 1200 model was purported to achieve a fillrate of 1.2 billion texels per second, with a geometry throughput of 15 million triangles per second. Most importantly, the card was originally claimed to achieve over 200 frames per second in id Software's *Quake III Arena* at maximum visual quality.*[2]

The 1200 model's claimed specifications would place it as the rough equivalent of the GeForce FX 5200 Ultra or Radeon 9200 Pro (very low performance GPUs of 2002 vintage), while its claimed performance would place it at the same level as the GeForce 3 Ti 500 or Radeon 8500 (high-end GPUs from 2000–2001). To compound matters, the cards' specifications were later updated to nearly double their original performance levels.

While the Glaze3D 1200 was supposed to achieve unheard of performance in video games, it was claimed that the 2400 and 4800 models would each be substantially more power-

ful in turn. Using two and four GPU configurations respectively, and including an additional geometry accelerator on the 4800, the higher-end Glaze3D cards were to be aimed at the very highest end of the video gaming market.*[2]

2.14.3   See also

- ATI Technologies
- NVIDIA

2.14.4   References

[1] Petri Nordlund. "Glaze3D" . Bitboys Oy.

[2] BitBoys Oy.   "BITBOYS OY UNVEILS GLAZE3D PRODUCT FAMILY" . Retrieved 2006-06-11.

2.14.5   External links

- Glaze3D Announced
- PDF version of a presentation by Petri Norlund, Chief Architect at BitBoys Oy in 1999.
- BitBoys at Siggraph - analysis of the Glaze3D cards.
- A Look Inside BitBoys - a detailed description of the development history of Glaze3D.

2.15   Godus

Godus is a god game style video game under development by the independent company 22Cans. The company launched a Kickstarter campaign to raise funds, and they met their funding goal of £450,000 (US$732,510) on 20 December 2012.*[1] *Godus* is designed by Peter Molyneux and is described by him as the spiritual successor to his earlier creation, *Populous*.

2.15.1   Gameplay

The player starts out by saving a man and a woman from drowning. Once the player leads them to the "Promised Land" , they will settle down and build a tent. They will "Breed" a worker, who will build another tent to live in. By using this strategy, the player will explore the world and improve your population through the ages. The main feature of this game is that the player is able to redesign the land levels at will. Different levels need more "Belief" than usual. The player will be able to explore at least one other world after finding a certain ship and gathering enough resources to repair it.*[2]

2.15.2 Development

Godus was the efforts of game designer Peter Molyneux and his development studio 22Cans.*[3] The game is the spiritual successor to Molyneux's earlier creation, *Populous*,*[4] and is inspired by his other titles: *Dungeon Keeper* and *Black & White*.*[5] Molyneux left his position at Microsoft in March 2012 to found 22Cans.*[6] With a staff of 20 people, the studio released its first game, *Curiosity – What's Inside the Cube?*, on 6 November 2012, and began working on *Godus*.*[6]

The company launched a Kickstarter campaign to help crowdfund the costs of producing the game, and the campaign met its funding goal of £450,000 (US$732,510) on 20 December 2012.*[1] Although the game was only funded two days before the campaign ended, any remaining pledges would be put towards stretch goals which would add features to the game such as more single-player modes, a cooperative mode, and adding support for Linux and the Ouya platform.*[1] 22Cans planned to release a prototype of the game on 13 December 2012, in an effort to attract more backers to the campaign.*[3] At the close of the campaign, £526,563 was raised and five out of the six stretch goals were reached, failing to achieve the goal for Ouya support.*[5] As of 21 December 2012, Molyneux had not set a release date for the game.*[4] As of the 13th September 2013, *Godus* was released on Steam early access as a beta version.*[7]

A freemium iOS version of the game was released 7 August 2014,*[8] and the Android version was released on 27 November 2014.*[9]

With the end of Molyneux's social experiment, *Curiosity – What's Inside the Cube?*, it was revealed that in *Godus* a single person will reign as the virtual god over all other players. It was also revealed that this same player will receive a portion of the revenue made by the game.*[10] As of February 2015, the winner Bryan Henderson has still not received the prize.*[11]*[12]

As of February 2015, the 22Cans staff had been reduced to a few people, likely indicating that some of the Kickstarter's goals such as multiplayer and a persistent world would not be met.*[13] Due to Molyneux seeming abandonment of the project, contrary to his earlier enthusiasm, many fans requested refunds. Furthermore, fans frequently cite "disgraceful treatment" and blatant lying in association with *Godus*, and by extension Molyneux.*[14]

2.15.3 Controversy

Eurogamer's Tom Bramwell expressed concern that successful game designers were funding their projects using Kickstarter, stating: "I'm looking at Kickstarter through the prism of Molyneux and Braben and Schafer and Fargo reaching out of their mansions and rattling their golden cups in my direction. They instantly put me in the mentality of a consumer [...] weighing up a pre-order against the potential fiction of their oft-broken pre-release promises. [...] It's not wrong because they are taking advantage of people – which may or may not be the case – but because this is absolutely not what Kickstarter is about." *[15] *PC Gamer* expressed similar concern: "one wonders if Molyneux couldn't have handled his own funding".*[16] Molyneux responded, saying, "I don't see why I, with my background, should be precluded from [Kickstarter]. I made the choice when I left Microsoft to become a small developer again and to define myself like a small developer defines itself, and that is someone who takes unbelievable risks – foolish risks like releasing Curiosity and doing Kickstarter." *[6] Molyneux also said that he invested a lot of his own money into the development studio 22cans.*[6]

Criticism has also been levelled at 22cans for their plan*[17] for a SteamOS and Linux version of Godus, despite being built on Marmalade*[18] which does not support Linux.

Additional criticisms were made over the freemium model chosen for the iOS version.*[19]

2.15.4 References

[1] Orland, Kyle (20 December 2012). "Project Godus reaches Kickstarter goal just under the wire" . *Ars Technica*. Retrieved 26 December 2012.

[2] "Proof of Playership Pictures" . *TinyPic*. Retrieved 10 May 2015.

[3] Orland, Kyle (13 December 2012). "Molyneux: Project Godus to release playable prototype tomorrow" . *Ars Technica*. Retrieved 26 December 2012.

[4] Sliwinski, Alexander (21 December 2012). "Godus Kickstarter concludes at £526K" . *Joystiq*. Retrieved 26 December 2012.

[5] Jackson, Mike (21 December 2012). "Godus Kickstarter closes at £526,563". *Computer and Video Games*. Retrieved 26 December 2012.

[6] Yin-Poole, Wesley (20 December 2012). "With Project Godus funded, Peter Molyneux can finally get some sleep" . *Eurogamer*. Retrieved 26 December 2012.

[7] Grubb, Jeffrey (13 September 2013). "Peter Molyneux's Godus beta available now on Steam Early Access" . *GamesBeat*. Retrieved 15 September 2013.

[8] "Godus – the Regenesis of the God Game Now Available Worldwide on iOS" . *Kickstarter*. 7 August 2014. Retrieved 7 August 2014.

[9] "Godus Now Available to Download on Android". *22cans*. 27 November 2014. Retrieved 28 November 2014.

[10] Younger, Paul (26 May 2013). "Edinburgh Curiosity winner reigns in GODUS, takes percentage of earnings". *PC Invasion*. Retrieved 9 July 2015.

[11] Yin-Poole, Wesley (11 February 2015). "The God who Peter Molyneux forgot". *Eurogamer*. Retrieved 9 July 2015.

[12] Walker, John (11 February 2015). "Loss Of Faith: Will Godus Ever Have A God Of Gods?". *Rock, Paper, Shotgun*. Retrieved 10 May 2015.

[13] Duwell, Ron (11 February 2015). "Peter Molyneux apologizes for Godus, promises a fix from new team". *TechnoBuffalo*. Retrieved 10 May 2015.

[14] Campbell, Colin (11 February 2015). "Peter Molyneux's Godus is a failure of trust, and a warning for others". *Polygon*. Retrieved 9 July 2015.

[15] Bramwell, Tom (15 December 2012). "Are the rich old men ruining Kickstarter?". *Eurogamer*. Retrieved 26 December 2012.

[16] "God Comes to Kickstarter: Peter Molyneux Pitches Populous Successor". *PC Gamer (UK)*: 9. January 2013.

[17] "22 Cans plan Linux version". Retrieved 4 December 2014.

[18] "Godus Made with Marmalade". Retrieved 17 November 2014.

[19] Hodapp, Eli (8 August 2014). "The Internet is Not Too Happy About Peter Molyneux's 'Godus'". *Touch Arcade*. Retrieved 10 August 2014.

2.15.5 External links

- Official website
- *Project GODUS* at Kickstarter

2.16 The Grinder

This article is about the video game. For the American television series, see The Grinder (TV series). For the snooker player, see Cliff Thorburn. For the poker player, see Michael Mizrachi.

The Grinder is a first person shooter video game for the Wii,[2] and a top-down shooter[3] for the PlayStation 3, Xbox 360, and Microsoft Windows by High Voltage Software.[4] It missed its scheduled release date and High Voltage Software have yet to make a statement regarding its status.[5]

2.16.1 Plot

In an alternate universe, mythical creatures such as vampires, werewolves and zombies are a reality, and the people of Earth have adapted to their presence. A small group of monster hunters, comprising Hector, Doc, Miko and AJ, are hired by the mysterious organization called "Book" to exterminate these threats while investigating the source of the outbreak.

2.16.2 Gameplay

In *The Grinder*, players control one of four characters, each with their own unique attributes which can be upgraded as the game progresses. Players will have to use different techniques to defeat certain foes, and will have access to a large number of weapons, some of which can be dual wielded. The game will also offer players complete customization of its controls, and is stated to support Wii MotionPlus.

An online multiplayer mode will be included in *The Grinder*, and each level can be played online cooperatively, with support of up to four players simultaneously, and will have multiple routes.[6]

At the 2010 Game Developers Conference, it was revealed that the Xbox 360 and PlayStation 3 versions of the game would be a top-down shooter, rather than a first-person shooter.

Characters

- **Hector**: A Mexican bounty hunter who has tracked targets on both sides of the border. Hector is boastful, arrogant and greedy. He used to be a professional hunter, but has annoyed so many people over the years that he cannot get a job with any of the professional companies.

- **Doc**: He wants to figure out what makes the monsters tick. Doc was a hunter for years but he left the life behind years ago when a mission went very badly. Since then, he has been working as an underground doctor.

- **Miko**: Miko is a Japanese assassin looking for a new challenge. She was getting bored with her job, until the night one of her targets turned out to be a werewolf. Despite nearly getting killed, she found the battle exhilarating and immediately set out to take down more monsters.

- **AJ**: An urban explorer who had a very bad experience and is looking for revenge. AJ is the sole survivor of a Slasher attack. She knows that it is only a matter of time before the unkillable monster that slaughtered her sorority sisters on a weekend camping trip catches up with her to finish the job.

- **Book**: The secretive contact responsible for hiring the four hunters to control the outbreak.*[6]

2.16.3 Development

The engine used for Wii edition of The Grinder, *Quantum3, allows a larger number of characters to be effectively drawn onscreen than most Wii titles to date.*

The Grinder will use the Quantum3 engine, which High Voltage confirmed would help push the amount of characters onscreen to the maximum. The developers are also working to have the game contain no loading times.*[6] High Voltage Software previously worked on *Hunter: The Reckoning* and confirmed that *The Grinder* was influenced by that game. *Left 4 Dead* was also an inspiration, as were grindhouse horror films.

The Grinder was in development for the Wii only, but a later interview revealed that ports will be made for other platforms.*[7] While the Wii version of the game is a first person shooter,*[2] the ported versions are top down shooters.*[3]

2.16.4 References

[1] Interview by Aaron Kaluszka. "Interview with High Voltage Software's Josh Olson". *nintendoworldreport.com*. Retrieved 2010-03-21.

[2] The Grinder staying FPS on Wii

[3] "Screenshot of HD version of The Grinder". Gamesblog.it. 2010-02-15. Retrieved 2010-02-15. |first1= missing |last1= in Authors list (help)

[4] Casamassina, Matt (2010-02-03). "The Grinder Goes Multiplatform". IGN. Retrieved 2010-02-04.

[5] http://games.ign.com/articles/995/995409p2.html#9

[6] Casamassina, Matt (2009-05-26). "IGN Dev interview". IGN. Retrieved 2009-05-27.

[7] The Grinder no longer Wii-exclusive, new trailer, due out Halloween 2011, HVS talks The Conduit sales, Go Nintendo

2.16.5 External links

- High Voltage Software's official website

2.17 Haystack (software)

Haystack was a never-completed program intended for network traffic obfuscation and encryption. It was promoted as a tool to circumvent internet censorship in Iran.*[1] Shortly after the release of the first test version, reviewers concluded the software didn't live up to promises made about its functionality and security, and would leave its users' computers more vulnerable.

2.17.1 History

Haystack was announced in the context of the perceived wave of Internet activism during 2009 Iranian election protests. There was a great deal of hype surrounding the Haystack project. The BBC's *Virtual Revolution* television series featured the software in the context of attempts to bypass network blocking software in Iran.*[2] The project was composed of a single programmer, and spokesperson, Austin Heap, who claimed to be a software developer based in San Francisco, California, but in fact Austin Heap was never a software developer, he was an unemployed marketing trainee . Austin named themselves as the Censorship Research Center.*[3]*[4]*[5] Early on in the project, Heap claims to have received a manual describing Iran's filtering software, written in Persian, from an Iranian official.*[6]

First person to raise the alarms was the Iranian blogger, Potkin Azarmehr, who claimed Haystack, much like the "emperor's clothes" never existed and no one was using such a software inside Iran to circumvent the internet censorship. Azarmehr also says the US State Department was so embarrassed after having given export exemption for technology to Iran for Haystack that they tried a damage limitation exercise by saying it had "security issues", whereas it never existed.

Azarmehr finally reported the matter to Evgeny Morozov, a technology journalist, and passed him a Beta version of the product, after he had enough of media hype about something that never existed. Morozov's article then led to the wind down of the project and its board members resigned in embarrassment. Haystack 'anti-censorship' software withdrawn over security concerns

Amidst criticism from technologists, including Jacob Appelbaum and Danny O'Brien, on September 13, 2010, the Washington Post reported[*][7] that security concerns had led to suspension of testing of Haystack. A message on the front page of the Haystack web site posted the same day confirmed the report, saying "We have halted ongoing testing of Haystack in Iran pending a security review. If you have a copy of the test program, please refrain from using it." The following day the BBC reported the same news and quoted Austin Heap of the CRC as stating that source code to the application would be released.[*][8]

2.17.2 Shutdown

The resignation of the only programmer on the project, Daniel Colascione, effectively ended development of the Haystack project.[*][9] The project web site is now defunct.

2.17.3 References

[1] "haystack: a project for iran" . Retrieved 2010-02-08.

[2] "The web makes the personal political". BBC News. 2010-02-05. Retrieved 2010-02-09.

[3] "Needles in a Haystack" . Newsweek. 2010-08-06. Retrieved 2010-08-06.

[4] "What's monitored online?". Tehran Bureau. 2010-01-18. Retrieved 2010-02-08.

[5] *The Virtual Revolution*. BBC. 2010-02-08. Retrieved 2010-02-08.

[6] William J. Dobson (2010-08-06). "Needles in a Haystack" . Newsweek. Retrieved 2010-08-07.

[7] Washington Post reports suspension of testing, retrieved September 13, 2010

[8] "Anti-censorship program Haystack withdrawn" . *BBC News* (BBC). 14 September 2010. Retrieved 15 September 2010.

[9] Award-Winning Haystack Security System Could Risk Iranian Lives, retrieved 2010-09-17.

2.17.4 External links

- Haystack website defunct
- Haystack on Twitter defunct
- Censorship Research Center defunct
- Cryptography, Iran and America: Worse than useless

2.18 Hellraid

Hellraid is an upcoming video game in development by Techland. It was scheduled to be released for Microsoft Windows, PlayStation 4 and Xbox One in 2015, but has been put on hold.

2.18.1 Gameplay

Hellraid will feature both co-operative and single player modes, as well as an online mode called 'Game Master'.[*][2] The game will also feature randomly generated content.[*][3] The co-op mode will be able to feature up to four players.[*][4] The game will also be played from a first-person perspective.[*][5]

2.18.2 Plot

The player will control Aiden (Nolan North), the last member of a cursed kin, who has formed an alliance with an old mage. Together, Aiden and the old mage must try to stop 'the infernal forces'.[*][6]

2.18.3 Development

The game was originally going to be a mod for *Dead Island*, which was also developed by Techland, before it became an individual game.[*][2] The game's working title was *Project Hell*.[*][3] The game was first announced in a press release on April 29, 2013. The press release stated that the game would be released in 2013.[*][7] The game was described as a mix "the best aspects of" *Dead Island* and *The Elder Scrolls* and a spiritual successor to the 1990s games such as *Hexen* and *Witchaven*.[*][8][*][9] In May 2014, Hellraid was delayed to allow it to be rebuilt on Techland's latest version of their in-house engine, Chrome Engine 6. Also in May 2014, Techland announced it was transitioning *Hellraid* to the PlayStation 4 and Xbox One. Along with this, Techland announced that it would be a digital-only game.[*][10] In May 2015, Techland announced that the company have put the development of *Hellraid* on hold as the game failed to meet the company's expectation and that the company wanted to focus on expanding the universe of the *Dying Light*, which was also developed by Techland and was released in January 2015.[*][11][*][12][*][13]

2.18.4 *Hellraid: The Escape*

Hellraid: The Escape is a video game set within the *Hellraid* universe.[*][14] Unlike *Hellraid*, *The Escape* puts emphasis on puzzle-solving instead of action.[*][15] It was developed by Shortbreak Studios with assistance from Techland and was released on iOS platform on 15 May, 2014.*[16] The game was powered by Epic Games' Unreal Engine.*[17] Upon release, the game received generally positive reviews. Aggregating review websites Metacritic have the game 75/100.*[18] TorchArcade gave the game a 4.5/5, calling it "a great experience on iOS and a rather successful combination of first-person exploration and puzzle elements." *[17] No in-apps purchase is available in the game and all post-launch content will be free to players.*[19]

2.18.5 References

[1] "Hellraid" . *uk.ign.com/*. Retrieved August 5, 2013.

[2] Connor Sheridan (2013-04-29). "First-person slasher Hellraid coming from Dead Island dev" . GamesRadar. Retrieved 2013-07-24.

[3] Gallegos, Anthony. "Dead Island Devs Announce Hellraid for 360, PS3 and PC" . IGN.

[4] Phillips, Tom. "Dead Island dev announces co-op hack-and-slash Hellraid" . Eurogamer.

[5] Makuch, Eddie. "Dead Island dev reveals Hellraid" . GameSpot.

[6] "Hellraid's E3 2014 Gameplay with Dev Commentary" .

[7] Hinkle, David. "Dead Island dev Techland's 'Project Hell' now Hellraid, out this year" . Joystiq.

[8] "Hellraid - Duchowy następca Hexen, Heretic i Witchaven" . PurePC.pl. Retrieved 2013-07-24.

[9] "Hellraid E3 Trailer « GamingBolt.com: Video Game News, Reviews, Previews and Blog". Gamingbolt.com. Retrieved 2013-07-24.

[10] Cook, Dave (April 30, 2014). "Hellraid re-announced for PC, PS4 and Xbox One following overhaul – new screens" . *VG 247*. Retrieved May 20, 2015.

[11] Crecente, Brian (May 21, 2015). "Techland's Hellraid sent back to drawing board, didn't meet expectations" . *Polygon*. Retrieved May 21, 2015.

[12] Phillips, Tom (May 21, 2015). "Dead Island dev dumps Hellraid development" . *Eurogamer*. Retrieved May 20, 2015.

[13] Futter, Mike (May 21, 2015). "It's A Cold Day, As Hellraid Development Put On Ice" . *Game Informer*. Retrieved May 21, 2015.

[14] Nunneley, Stephany (1 April 2014). "Hellraid: The Escape release date announced for iOS" . *VG 247*. Retrieved 31 March 2015.

[15] "First Person Game Hellraid: The Escape Announced For iOS" . *Touchapplay*. 2 April 2014. Retrieved 31 March 2015.

[16] Farokhmanesh, Megan (1 April 2014). "Hellraid: The Escape heading to iOS May 15" . *Polygon*. Retrieved March 31, 2015.

[17] Ford, Eric (9 July 2014). "Hellraid: The Escape review: First person puzzling at its Finest". *TouchArcade*. Retrieved 31 May 2014.

[18] "Hellraid: The Escape for iOS reviews" . *Metacritic*. CBS Interactive. Retrieved March 31, 2015.

[19] Watts, Steve (2 April 2014). "Hellraid: The Escape announced for iOS, coming in May" . *Shacknews*. Retrieved 31 March 2015.

2.18.6 External links

- Official website

2.19 Highlander: The Game

Highlander: The Game is a cancelled action role-playing game based on the *Highlander franchise*; it was to be published by Square Enix for Windows, PlayStation 3, and Xbox 360.*[2]*[3] It was announced on January 14, 2008 by way of a trailer on Gametrailers.com. On December 10, 2010, it was announced that the game has been officially canceled.*[4]

2.19.1 Gameplay

Much of the gameplay would revolve around sword battles with other Immortals

Highlander: The Game was to be a third-person action game featuring a previously unknown Immortal, Owen MacLeod as the main character.*[5]

An Immortal can only be killed via decapitation. Thus, Owen cannot die from other wounds that would be fatal to a mortal such as gunshots, electrocution, and falls from a great height. At most, he can "die" for a brief period of time, reviving fully intact once his body has healed itself. As such, certain encounters would have involved Owen deliberately performing incredibly dangerous actions, such as using his body as a conductor for high-voltage electrical currents and jumping off of buildings to escape pursuers.

Over and above his ability to withstand otherwise lethal attacks, Owen would be able to use certain magical abilities, going against the franchise's previous depictions of Immortals as ordinary men and women who do not age. Such abilities, which were never explained in depth, included "chi balance", "fireblade", "wind fury" and "stone armor", to name a few. Like any other Immortal in the Highlander franchise, Owen must wield various types of swords to do battle with other Immortals. Owen can also choose between different combat styles to defeat his enemies.

The game was to span roughly eighteen missions in which Owen would encounter up to seventy-seven unique characters. Some of these characters would have been Immortals just like Owen and must be slain. Some of these characters were previously known - Connor MacLeod and Duncan MacLeod were both expected to make appearances and Methos was confirmed as Owen's mentor.*[6]

2.19.2 Story

Owen is forced to compete as a gladiator in ancient Rome

The story, written by David Abramowitz who was Creative Consultant (Head Writer) on *Highlander: The Series*,*[1] revolves around Owen MacLeod, ancestor to both Connor MacLeod and Duncan MacLeod. It begins over two thousand years ago. And similar to other *Highlander* films and media, the story spans several timelines including Ancient Gaul, Pompeii, and finally modern-day New York City.*[7]

Owen is captured and enslaved by Romans who force him to compete as a gladiator. During this time, Owen dies only to come back to life. Methos, the oldest living Immortal, approaches Owen to be his mentor. He teaches Owen about The Game and how he and other immortals can only be slain by beheading.

As with other Immortal MacLeods, Owen is pursued throughout his life by a nemesis. This enemy proves to be extremely powerful, one that Owen is unable to defeat. Owen learns of a magical stone, fragments of which are scattered all over the world. Throughout the game, Owen embarks upon a quest to recover these fragments and restore the stone in an attempt to gain the power to overcome his foe.*[1]*[8]

2.19.3 See also

- Highlander (video game)
- Highlander: The Last of the MacLeods

2.19.4 References

[1] "Eidos Unleashes 'Highlander' Game". *Wireless News* (M2 Communications, Ltd.). 2008-01-21.

[2] Ward, David (2008-01-18). "Eidos announces 'Highlander' game". *The Hollywood Reporter* (Nielsen Business Media).

[3] "Eidos Develops *Highlander* Game". *SciFi.com*. 2008-01-18. Archived from the original on 2007-11-15. Retrieved 2008-11-12.

[4] http://ps3.ign.com/articles/113/1139727p1.html

[5] "*Highlander*: Immortal Combat," *GamePro* 235 (April 2008): 21.

[6] "Fresh Highlander details pop up". Eurogamer. 2008-01-18. Retrieved 2008-02-06.

[7] "Highlander History," *GamePro* 235 (April 2008): 21.

[8] "GC: Highlander details emerge". Eurogamer. 2007-08-24. Retrieved 2008-02-06.

2.19.5 External links

- Game Trailer
- Preview on Geek.com

2.20 Holographic Versatile Disc

The **Holographic Versatile Disc (HVD)** is an optical disc technology developed between April 2004 and mid-2008 that can store up to several terabytes of data on an optical disc 10 cm or 12 cm in diameter. The reduced radius reduces cost and materials used. It employs a technique known as collinear holography, whereby a green and red laser beam are collimated in a single beam. The green laser reads data encoded as laser interference fringes from a holographic layer near the top of the disc. A red laser is used as the reference beam to read servoinformation from a regular CD-style aluminium layer near the bottom. Servoinformation is used to monitor the position of the read head over the disc, similar to the head, track, and sector information on a conventional hard disk drive. On a CD or DVD this servoinformation is interspersed among the data. A dichroic mirror layer between the holographic data and the servo data reflects the green laser while letting the red laser pass through. This prevents interference from refraction of the green laser off the servo data pits and is an advance over past holographic storage media, which either experienced too much interference, or lacked the servo data entirely, making them incompatible with current CD and DVD drive technology.[*][1]

Standards for 100 GB read-only holographic discs and 200 GB recordable cartridges were published by ECMA in 2007,[*][2][*][3] but no holographic disc product has appeared in the market. A number of release dates were announced, all since passed.[*][4]

2.20.1 Technology

Current optical storage saves one bit per pulse, and the HVD alliance hopes to improve this efficiency with capabilities of around 60,000 bits per pulse in an inverted, truncated cone shape that has a 200 μm diameter at the bottom and a 500 μm diameter at the top. High densities are possible by moving these closer on the tracks: 100 GB at 18 μm separation, 200 GB at 13 μm, 500 GB at 8 μm, and most demonstrated of 5 TB for 3 μm on a 10 cm disc.

The system uses a green laser, with an output power of 1 watt which is high power for a consumer device laser. Possible solutions include improving the sensitivity of the polymer used, or developing and commoditizing a laser capable of higher power output while being suitable for a consumer unit.

2.20.2 Competing technologies

HVD is not the only technology in high-capacity, optical storage media. InPhase Technologies was developing a rival holographic format called Tapestry Media, which they claim will eventually store 1.6 TB with a data transfer rate of 120 MB/s, and several companies are developing TB-level discs based on 3D optical data storage technology. Such

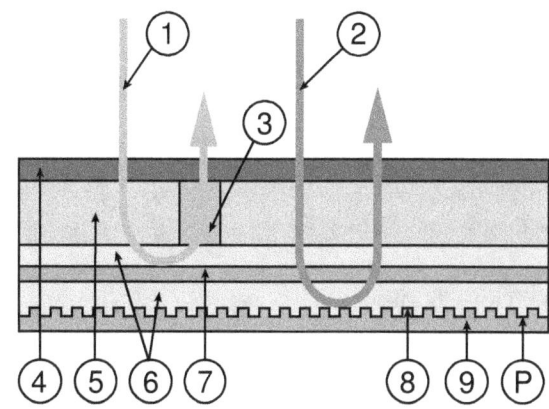

Holographic Versatile Disc structure
1. *Green writing/reading laser (532 nm)*
2. *Red positioning/addressing laser (650 nm)*
3. *Hologram (data)(shown here as brown)*
4. *Polycarbonate layer*
5. *Photopolymeric layer (data-containing layer)*
6. *Distance layers*
7. *Dichroic layer (reflecting green light)*
8. *Aluminium reflective layer (reflecting red light)*
9. *Transparent base*
P. *Pit pattern*
(Illustration is not to scale.)

large optical storage capacities compete favorably with the Blu-ray Disc format. However, holographic drives are projected to initially cost around US$15,000, and a single disc around US$120–180, although prices are expected to fall steadily.[*][5] Since InPhase Technologies were unable to deliver their promised product, they ran out of funds, and went bankrupt in 2010.[*][6]

2.20.3 Holography System Development Forum

The Holography System Development Forum (HSD Forum; formerly the HVD Alliance and the HVD FORUM) is a coalition of corporations purposed to provide an industry forum for testing and technical discussion of all aspects of HVD design and manufacturing.

As of March 2012, the following companies are members of the forum:[*][7]

- CBCGroup
- Daicel Chemical Industries ltd.
- FujiFilm
- Konica Minolta Inc.

- Kyoeisha

- Pulstec

- Shibaura Mechatronics Corporation

- Oracle Corporation

- Teijin Chemicals Ltd.

- Tokiwa Optical Corporation

The following are supporting companies of the forum as of March 2012:

- Kodate Laboratory

2.20.4   Standards

On December 9, 2004, at its 88th General Assembly, the standards body Ecma International created Technical Committee 44, dedicated to standardizing HVD formats based on Optware's technology.

On June 11, 2007, TC44 published the first two HVD standards:[*][8] ECMA-377,[*][2] defining a 200 GB HVD "recordable cartridge" and ECMA-378,[*][3] defining a 100 GB HVD-ROM disc. Its next stated goals are 30 GB HVD cards and submission of these standards to the International Organization for Standardization for ISO approval.[*][9]

2.20.5   References

[1]   "What's New" . 2004-08-23. Archived from the original on 2004-10-09.

[2]   "Information Interchange on Holographic Versatile Disc (HVD) Recordable Cartridges – Capacity: 200 Gbytes per Cartridge" . *ECMA-377*.

[3]   "Information Interchange on Read-Only Memory Holographic Versatile Disc (HVD-ROM) – Capacity: 100 Gbytes per disk" . *ECMA-378*.

[4]   "Maxell focuses on holographic storage". *CNET News.com*. 2005-11-28. Archived from the original on 2012-07-22. Retrieved 2007-05-28.

[5]   "Hitachi-Maxell to Ship Holographic Storage this Year" . *DailyTech*. 2006-08-03. Retrieved 2007-05-28.

[6]   "Holographic storage bites the dust" .

[7]   http://web.archive.org/web/20120313203810/http: //hvd-forum.org/forum/members.html

[8]   "Ecma releases new Holographic Information Storage Standards" . *Ecma press release*. 2007-07-04.

[9]   "Ecma standardizes Holographic Information Storage" (PDF). *Ecma press release*. 2005-01-26.

2.20.6   External links

- DaTARIUS signs agreement with InPhase Technologies to be their sole sales, service and support supplier of Tapestry Media hardware and media to ship starting in 2007 (300 GB WORM discs) with 600 GB discs and re-writable technology in 2008 as well as 1.6 TB media available in 2010.

- Optware, creator of HVD format.

- InPhase, a now bankrupt, company that developed a competing holographic storage format.

- Video explaining holographic storage – PC Magazine, October 4, 2006

- Holography system rides single beam EE Times, February 27, 2006 – interview with Hideyoshi Horimai and Yoshio Aoki of Optware Corp.

- Holographic storage standards eyed EE Times, February 28, 2006 – article about the upcoming technical committee meeting to begin standardization of HVD.

- How stuff works explains how HVD works.

- Elusive Green Laser Is Missing Ingredient Wall Street Journal February 13, 2008

- General Electric unveils 500GB optical disc storage

2.21   Huxley (video game)

Huxley (Korean: 헉슬리) is a multiplayer first-person shooter computer game with persistent player characters published by Webzen Games Inc.. It is being developed for Microsoft Windows. A Xbox 360 port was planned, but it has been put on an indeterminate hold.[*][2] *Huxley* initially was going to be cross platform, but according to statements made at the 2009 E3 Expo press conference that feature is currently excluded from development.[*][3] The contract to operate the game in China was sold to The9 for $35 million USD on February 12, 2007, considered the largest export transaction to date for a Korean-developed game.[*][4]

A trailer for the game was released in 2007 as a special DVD used to demonstrate LG LCD TVs.

In June 2009 NHN USA released the first English Closed Beta Test via its free games portal ijji.com. The initial test had a small number of users and was carried out over a space of two weeks. Keys for the test were made available through ijji (Globally) and FilePlanet (USA and Canada).

The second Closed Beta Test was initiated in late-July 2009 and lasted until August 12. The second test allowed many more players to test the game. During the last two days of the test a high-volume stress test was carried out on to the servers where everyone with an ijji account was permitted to play the game during test hours.

In April 2010 Huxley was integrated with Hangame game portal*[5] and went into open beta on May 3.*[6]

In August 2010 an official message was posted on the North American Huxley forums at IJJI. Huxley for the North American region will now be self hosted by Webzen.

On December 30, 2010, the Korean service for *Huxley* was discontinued.

2.21.1 Plot

In the near future, Nuclearites bombarded the world. Destructive earthquakes, massive tsunami and dramatic climate changes wreak havoc around the globe, isolating continents and driving the human race into chaos. Those who survive the destruction dream of tranquillity, but an eruption among the human race and the appearance of horrible mutants drives the world into further disorder. Racism and oppression cause rebellious uprisings and war that divide the landscape between two powers: Sapiens and Alternative. At the heart of the war emerges a powerful energy source called the Lunarites. The Lunarites were created by Huxley, a scientist and possible saviour.

Both factions seek glory and victory, fighting mercilessly for the Lunarites and their very existence.

The story was thought to be based on the novel *Brave New World* by Aldous Huxley, hence the name, however Webzen has stated there are no tie-ins to the book's story saying it was just an inspiration.*[7]

The game's visual style is reminiscent of the *StarCraft* series of games, which are extremely popular in South Korea (where the developer is based).

2.21.2 Gameplay

The speed and style of *Huxley* is fast paced and team oriented, combining gameplay from twitch shooters such as *Unreal Tournament* with the character advancement and large worlds seen in MMORPG's like *World of Warcraft*. In a recent interview, *Huxley's* main producer Kijong Kang said that the cities in *Huxley* will be able to accommodate up to 5,000 (according to recent publications that number

was increased to 10,000 since summer 2006) people, and the individual battles will support over a hundred players. However, due to major changes in the game's engine and budget, the number of battling players has been reduced to 64 players (32 against 32) at the most. This number may or may not increase once the game is released.

There are two main types of Player versus Player battles. These are called *Skirmishes* and *Battlefields*. The skirmishes are small battles between two teams of 8 players on small to medium sized maps. One player acts as the host player and can choose between a variety of settings and gametypes. The other players in the game then connect to the host and are separated into teams based on their character's faction. Battlefields are much larger scale battles with higher stakes and more players. In the *Open Frontier Test* of *Huxley*, this mode supported 64 players, 32 of each faction. Battlefields generally take place on much larger maps, and have objectives that players need to capture, retrieve, gather, or defend. Because hosting 64 players on a player's PC would be incredibly difficult, the Battlefields are hosted by dedicated servers.

In the Player versus Environment portion of *Huxley*, players can either play alone or group up with three others to kill AI-controlled monsters and complete quests. All fighting takes place in instance dungeons. There is also a town where players can meet up to talk, trade, shop, and form parties.

2.21.3 Leveling and experience

In the early part of the character advancement player system, players can shape their character in the style they like best. After that, players can then add depth to their characters. 'Experiences' and 'Battle points' are two elements of character advancement. 'Experiences' will affect the earlier part of character development and 'Battle Points' will affect the later part of a character's development in a big way.

In the earlier part, by acquiring licenses, characters can have opportunities to use upgraded weapons and armor. In the later part of the game, players will concentrate on developing their characters to be more effective under any circumstance. One developer has stated, "We are planning to make the earlier part of character development relatively fast and the later part of development a bit slower but more abundant. This is because we decided that too much difference between characters abilities that affect combat result is not good for an FPS game."

Webzen has considered the fact that in most MMOGs the players at earlier levels have no chance of defeating those at the higher levels, and therefore they have adjusted the game to make skill more significant than long periods of playing the game and leveling up. Leveling up will give players advantages, such as more slots for upgrades and perhaps faster aiming, but a lower level player can still measure up to a higher one.

2.21.4 Vehicles

Huxley includes vehicles for both player-vs-player and player-vs-environment gameplay, as well as general transportation. Gameplay footage has shown several visually interesting SciFi vehicles, including an APC, an aircraft resembling *Unreal Tournament 2004*'s Raptor, a large energy based tank, and a small assault buggy. Vehicles are used in both combat and transportation.

Developer interviews have also stated that players can obtain motorcycles for personal transportation, and that they will mainly serve as status symbols for wealthy players and as a quicker mean of transportation through town.

2.21.5 Weapons

Huxley has nine different types of weapons. Each class specializes in three weapons, and also has the option of using four weapons of the other classes, although with less proficiency.

There are also several different modifications of each type of weapon. These usually have minor tradeoffs like doing more damage with a lower fire rate, or being more accurate with less damage.

Some special weapon types start to appear for higher level classes. Notable ones of these are a splitting rocket launcher in which the rocket splits into two after a distance and an optical rifle which does a damage over time or slows the enemy. For the Enforcer they get special Flingers which have a longer distance of fire and will either stick the to wall or be able to be bounced off the wall. The Phantom gets Sniper rifles which have a chance to cause a devastating damage over time.

2.21.6 Abilities

Abilities, or skills, can be used in battle to give a player to a tactical advantage. As a player levels up, they are rewarded with new and more powerful skills. The type of skills available to the player differs based on their class. Skills are used by being "equipped" to armour. The number of skills that can be equipped depends on the players level and the rank of the armour. This system is very similar in concept to *Call of Duty 4*'s perk system.

There are two types of skills, active and passive. Passive skills are skills that act as buffs, or skills that activate automatically when a certain condition is met, examples of this include health regeneration, and the ability to drop a flash grenade when low on health. Active skills are activated manually by hitting the assigned key.

2.21.7 Hosting

In June 2008, NHN USA announced that it had secured the rights to distribute *Huxley* via its ijji portal.[*][8][*][9]

In August 2010, NHN USA announced that they will be transferring the publishing duties of *Huxley: The Dystopia* to developer Webzen.[*][10][*][11]

2.21.8 Soundtrack

The soundtrack for *Huxley* was composed by Kevin Riepl.[*][12] The score was recorded by the 80-piece Hollywood Studio Symphony at Warner Bros in March 2007.[*][13][*][14]

Track listing

2.21.9 See also

- MMOFPS

2.21.10 References

[1] Webzen Games (4 October 2007). "Acclaimed composer Kevin Riepl scores Webzen's MMOFPS Huxley". Webzen Games Inc. Retrieved 2014-08-15.

[2] *GDC 09: A First Look at Huxley*

[3] *Huxley* Interview (Shacknews)

[4] Webzen Inc. Webzen sells China rights for *Huxley* to The9 February 12, 2007.

[5] "한게임서비스변경안내". Hangame. April 20, 2010. Retrieved May 16, 2010.

[6] "헉슬리: 더디스토피아 Open Beta 서비스일정안내". Hangame. May 3, 2010. Retrieved May 16, 2010.

[7] Producer Kijong Kang talks about the online first-person shooter with amazing visuals and an unusual name (Yahoo Games)

[8] News: *Huxley* gets publisher, dated for 2008 - ComputerAndVideoGames.com

[9] Webzen and NHN USA executes *Huxley* licensing agreement for North America and Europe

[10] *Huxley: The Dystopia Returns to Webzen Hands.* mmofront.com, August 28, 2010. Retrieved August 29, 2010.

[11] *Huxley Official Statement.* posted in section *Announcements* of ijji.com forums, August 27, 2010. Retrieved August 29, 2010.

[12] IGN Music (5 October 2007). "Kevin Riepl Joins Huxley. MMOFPS gets the sonic treatment from Gears of War composer". *IGN*. Ziff Davis. Retrieved 18 October 2014.

[13] Goldwasser, Dan (19 March 2007). "Kevin Riepl Scores a Brave New World of Video Game Music in Huxley". Scoring Sessions. Retrieved 18 October 2014.

[14] Webzen Games (4 October 2007). "Acclaimed Composer Kevin Riepl Scores Webzen's MMOFPS Huxley". Webzen Games. Retrieved 18 October 2014.

2.21.11 External links

- Official Korean Website

2.22 Hyper CD-ROM

The **Hyper CD-ROM** is an optical data storage device similar to the CD-ROM with a multilayer 3D structure, invented by Romanian scientist Dr. Eugen Pavel.

2.22.1 Characteristics

The disc has a height of 1.2 mm and a diameter of 120 mm and can be produced with existing technology. The storage capacity of one such disk is 1,000,000 GB (1 PB), as storage occurs on 2 000 different levels layered inside the glass body of the disk.

- Capacity: 1 PB (with possibility of extension up to 100 EB)

- Medium transfer rate: 300 Mbit/s (may be more)

- Hard drive dimensions: 80 x 150 x 300 mm

- Disk dimensions: 1.2 mm ø 120 mm

- Temperature resistance: up to 550°C

- High reliability

- Maximum usage period: 5,000 years

In November 1999, the disk was presented at EUREKA "48th World exhibition of Innovation and New Technology" in Brussels.

The technology has gained recognition in 21 countries including the U.S., the EU, Canada, Japan and Israel.

There has not been any large-scale production of the Hyper CD-ROM, although several firms like IBM, Compaq, Philips and other Hollywood businesses have taken an interest in this form of data storage.

2.22.2 External links

- Storex Technologies, the Company producing the disc.

- Press Release, about 1PB optical disc.

2.23 Imperator Online

Imperator Online was an Alternate Earth MMORPG being developed by Mythic Entertainment, the makers of Dark Age of Camelot, one set in a future world where Ancient Rome never fell. Minor changes at important moments in Roman history create an extremely different timeline for Earth, leading to an interstellar Roman Republica and thousands of years of galactic Pax Romana. However, the game's production was canceled in July 2005, when Mark Jacobs announced that they had acquired the Warhammer Online license from Games Workshop.

2.24 Jet Thunder

Jet Thunder: Falklands/Malvinas is a combat flight simulation product in development by Thunder Works Studios LLC. It is aimed to be historically accurate (featuring the 1982 Falklands War) and will feature aircraft, campaign and missions for both sides of the conflict.[1]

The project was originally being conducted outside of the standard publisher driven model, by part-time enthusiasts located in 3 countries (Brazil, UK and Argentina). However, in October 2010, the developers signed a partnership with Aerosoft GmbH, a German publisher specialized in simulation products.[2]

2.24.1 Features

Jet Thunder: Falklands/Malvinas will be the first hardcore simulation to accurately portray the South Atlantic air war as an out of the box, full product.

The current plan under Aerosoft publishing is to release the project in stages, first having the British Harrier and the Argentinean FMA IA 58 Pucará as flyable (sold by download only), then adding the other Argentinean aircraft and airbases in a second download release, and later having a boxed/DVD release with all the features originally promised.[*][3]

In the full boxed/DVD release, the player flyable aircraft will be the A-4 Skyhawk, Sea Harrier, Mirage III and Super Etendard[*][4] and two campaigns will be available: British perspective and Argentine perspective, and the outcome of the campaign can be changed by effective player's actions (dynamic campaign).[*][1]

As of October, 2014 there still has been no release of a game or even a playable demo.

2.24.2 External links

- Official website

2.24.3 References

[1]   "Frequently Asked Questions about Jet Thunder" .

[2]   "News about Jet Thunder 001 (English)".

[3]   "Goose Greenlight: Jet Thunder Gets Publisher" .

[4]   "Jet Thunder features" .

2.25 Killing Day

Killing Day is a cancelled video game that was being developed for the PlayStation 3.[*][1] The game was first revealed during Sony's E3 press conference in 2005. However since then the game was subsequently cancelled in the middle of development. Although, in summer of 2009, developer Ubisoft filed for a trademark on *Killing Day*, sparking rumors that the game is still in development.[*][2] As of September 2012, no news of the game's development has been revealed. On January 4, 2013, a new trademark application was filed but there have been no new comments by Ubisoft.

2.25.1 References

[1]   http://www.gamespot.com/ps3/action/killingday/index.html

[2]   http://www.gamespot.com/ps3/action/killingday/news.html?sid=6210351&om_act=convert&om_clk=newsfeatures&tag=newsfeatures;title;1

2.26 Kingdom Under Fire II

Kingdom Under Fire II is a video game set in a high fantasy setting developed by Blueside which merge real-time strategy (RTS), role-playing game (RPG) and massively multiplayer online game (MMO) genres - the game is to have a single player, and online multiplayer mode. The game follows on chronologically from *Kingdom Under Fire: Circle of Doom*, and is the first RTS game set in the *Kingdom Under Fire* universe to be released since the 2005 *Kingdom Under Fire: Heroes*.

The game was announced in January 2008, and has been subject to delay and changes to release platforms; A closed beta-test began on December 2011 in South Korea.

In November 2013, the developers announced that a version for the PlayStation 4 was in development.

2.26.1 Gameplay

The game combined action RPG with real-time strategy. The player was to control a hero who can command various troops. The player had full control of their character and can run around freely with their troops following them. When controlling other units, the game played like a typical RTS; cannons for example could be used to break down castle walls and spearmen could be used to protect troops from cavalry.[*][3]

2.26.2 Plot

The plot of *Kingdom Under Fire II* comes chronologically after the events in *Circle of Doom* - which took place within the alternative dimension of the game's antagonist - Encablossa. The game's events continue the story of the *Kingdom Under Fire* universe, 150 years after *Kingdom Under Fire: The Crusaders*, and introduce a new faction - the 'Encablossians', who have been brought from the Encablossan dimension by Regnier,[*][1][*][4] an antagonist in previous games.

The game was to explore the wars between three factions: the Human Alliance, Dark Legion, and Encablossians in their struggle for control of the game's world, the continent of Bersia.[*][4]

2.26.3 Development

The game was originally announced in January 2008 to be published on PC and console platforms in 2009.[*][5] The game was to be a blend of MMO and RTS gameplay to create a "Multiplayer Online Action RTS" or MMORTS;

key elements of the gameplay design include a single player 'story mode' as in the earlier RTS *Kingdom Under Fire: The Crusaders*, and a separate online multiplayer element where players would be able to undertake battles, both castle sieges, and pitched battles in fields.[*][4] The MMO aspect of the game was to take place in a persistent world, with players able to form and join guilds and take part in territorial battles (in up to 15 vs 15 player matches).[*][1]

In July 2009, Blueside confirmed an Xbox 360 version for release in 2010, the Xbox 360 version was stated to have greater emphasis on the single player campaign, and to also have an online multiplayer mode, as well as downloadable content, but a MMORTS mode was not confirmed.[*][6][*][7][*][8]

In March 2010, the project was reported by MMOSITE.com to be rumoured to be in development hell, with developer Blueside in financial trouble, with some employees salaries unpaid, and also as being affected by the departure of part of Blueside's development team to form a new company. In an interview representatives from Blueside and Phantagram stated that 90 persons were working on the project, and that the game had been delayed due a shortage of manpower caused by simultaneous development of console and PC versions, as well as delays due to inexperience with online game development. The lead developer Lee Sang Yong stated that there was no real problem with the development.[*][9]

In October 2010, the Xbox 360 version was announced to be delayed to at least Christmas 2011 due to restrictions imposed on Xbox Live games by Microsoft relating to billing; development of the actual game system was said by the developers to be complete.[*][10] The projects budget had exceeded $20 million by late 2010.[*][10]

In November 2010, Blueside announced the development of a PlayStation 3 version, initially with a projected release date in 2012, citing the suitability of the PS3's online mechanism's for adding updates. Issues relating to the online Xbox 360 version were stated to be under continued discussion with Microsoft.[*][11]

The game was developed with a video game rating of 15+, though the developers expected it may receive an 18+ rating due to the combat and war elements.[*][12] A closed beta test began in late 2011.[*][13][*][14]

In November 2013, the developer Blueside announced the game would be released for PC in Korea in the next month, and that the game was also in development for the PlayStation 4.[*][15] At the G-Star Korean gaming industry convention (November 2013), producer Sang-Yoon Lee announced that the game would be released on PC in South Korea, Malaysia and Singapore in the next month.[*][16][*][17] At G-Star Sang-Yoon Lee apologised for

delays to the game, explaining that the development had been hampered by loss of half of its original 160 person team, as well as market conditions in South-East Asia, where the high computer specification requirements of the game meant a limited market for the product, requiring work to allow the game to run a relatively low powered hardware (e.g. GeForce 7600 graphics card); at the same time a PS4 version of the game was stated to be in development and expected to be complete by around May 2014.[*][17][*][18]

MMOGAsia announced open beta for *Kingdom Under Fire II* for May 22, 2014. The developer revealed that there will be no data wipes after the launch of open beta, so any progress made during the testing phase will carry over to full release.[*][19]

2.26.4 References

[1] "Kingdom Under Fire II Q&A: Bringing Massively Multiplayer to RTS", *uk.gamespot.com* (GameSpot), 28 February 2009: 1–2

[2] "Kingdom Under Fire II Slated For Spring 2016 On PlayStation 4". *Siliconera*. September 17, 2015.

[3] *Kingdom Under Fire II: The Human Alliance*, IGN, 18 October 2011

[4] Charles Onyett (3 March 2008), "Kingdom Under Fire 2 Q&A", *uk.pc.ign.com* (IGN)

[5] Robert Purchese (28 January 2008), "Kingdom Under Fire gets MMO sequel", *www.eurogamer.net* (Eurogamer)

[6] Blueside (press release) (8 July 2008), César A. Berardini, ed., "Kingdom Under Fire II Update + New Screenshots", *news.teamxbox.com* (Team Xbox)

[7] Oli Welsh (9 July 2009), "360 Kingdom Under Fire II not an MMO", *www.eurogamer.net* (Eurogamer)

[8] Tom Magrino (8 July 2009), "Kingdom Under Fire II igniting 360 in 2010", *uk.gamespot.com* (GameSpot)

[9] "Kingdom Under Fire II is Suffering an Ordeal or Goes to Miscarry", *news.mosite.com* (MMOsite.com), 2 March 2010: 1–2

[10] Robert Purchese (26 October 2010), "Xbox Live policy slows 360 KUFII release", *www.eurogamer.net* (Eurogamer)

[11] Robert Purchese (18 November 2010), "Kingdom Under Fire II heading to PS3", *www.eurogamer.net* (Eurogamer)

[12] "KUF2 Interview: The In-game War Probably Make It an 18+ MMOG", *news.mmosite.com* (MMOsite.com), 15 August 2011: 1–2

[13] Cindyhio, ed. (27 September 2011), "Kingdom Under Fire 2 to Kick off a CBT in Late 2011", *news.mmosite.com* (MMOsite.com): 1–2

[14] *Kingdom Under Fire II 1st CBT Simple Review - MMORPG News*, mmosite, 14 Dec 2011

[15] Scammel, David (12 Nov 2013), "Kingdom Under Fire 2 confirmed for PS4", *www.videogamer.com*

[16] "[GSTAR2013] " 다음달, 동남아부터 시작" ' 킹덤언더파이어 2' 이상윤 사장 인터뷰", *www.inven.co.kr* (in Korean), 17 Nov 2013

[17] Neltz, András (19 Nov 2013), "Kingdom Under Fire II Is Still on the Way. Honest!", *kotaku.com*

[18] 킹덤언더파이어 2, 내년 상반기에는 꼭 선보이겠다, *www.thisisgame.com* (in Korean), 18 Nov 2013

[19] "Kingdom Under Fire II Open Beta Launches This Week" . *MMO Attack*. 19 May 2014. Retrieved 20 May 2014.

2.26.5   External links

- Kingdom Under Fire ll MMOG website

- Kingdom Under Fire ll Korean website

- Kingdom Under Fire II official website

- "Blueside : Kingdom under Fire II", *www.blueside. co.kr* (Blueside)

2.27   Kirby's Return to Dream Land

Kirby's Return to Dream Land, known in Japan as *Hoshi no Kābī Wii* (星のカービィ Wii "Kirby of the Stars Wii") and in Europe and Australia as *Kirby's Adventure Wii*, is a *Kirby* video game and the twelfth platform installment of the series, developed by HAL Laboratory, and published by Nintendo. While *Kirby's Epic Yarn* was released in 2010, *Kirby's Return to Dream Land* is the first traditional *Kirby* platforming home console game since *Kirby 64: The Crystal Shards*, which was released in 2000 for the Nintendo 64. The title was first released in North America on October 24, 2011, and later in Europe on November 25, 2011.

Kirby's Return to Dream Land features the staple gameplay of traditional *Kirby* platform games, in which the eponymous character Kirby possesses the ability to inhale and copy enemies in order to attain forms which give him a variety of attacks such as breathing fire or swinging a sword. The game supports cooperative multiplayer gameplay, allowing up to four players to control various *Kirby* characters, including Waddle Dee, King Dedede, and Meta

Knight. *Kirby's* plot focuses on the characters retrieving the scattered pieces of a crashed alien spaceship.[8]

The game was first announced as a Nintendo GameCube title to be released in late 2005,[9] but development was later shifted to its successor console, the Wii. The game was presumed to be canceled until it was re-announced in 2011. It is notable as the first traditional Kirby game to be released on console in 11 years following *Kirby 64: The Crystal Shards*.

In the January 2015 Nintendo Direct, it was announced that *Kirby's Adventure Wii* and other Wii games will be released for download through the Wii U's Nintendo eShop. In Japan, *Kirby's Adventure Wii* was made available on January 28, 2015,[5] and was available in Europe, Australia and New Zealand on February 19, 2015 [6][7] and North America on July 30, 2015.

2.27.1   Gameplay

See also: Gameplay in the Kirby series

Kirby's Return to Dream Land is a 2.5D side-scrolling platform game, controlled by holding the Wii Remote sideways. Differing from other games in the *Kirby* series, this game features a completely 2.5-dimensional style of gameplay, and 3D modeled characters, enemies, mid-bosses and bosses are used, rather than sprites like has been done since *Kirby's Dreamland*. The main objective is to assist the alien Magolor, whose spaceship, the *Lor Starcutter*, has crashlanded on Pop Star. The player is tasked with collecting the scattered pieces of the spaceship, which are dispersed and hidden within the game's levels, each of which take place in various areas of the planet.[10]

Kirby's Return to Dream Land *supports cooperative multiplayer for up to four players.*

Kirby, the main protagonist, retains his signature ability to inhale indefinitely, allowing him to suck in nearby objects and enemies. The inhaled objects can either be propelled back out as projectiles or swallowed. Certain enemies, when swallowed, allow Kirby to extract their special qualities, giving Kirby access to a wide variety of different powers, called "copy abilities." Copy abilities are used to defeat stronger enemies and clear special environmental obstacles, though Kirby can only possess one copy ability at a time. Similar to *Kirby Super Star*, each copy ability has multiple forms of attack that are summoned depending on the player's button-presses. New powers introduced to this game are the Whip, which lets Kirby grab out of reach items, Water, which allows Kirby to cool fiery areas, and Leaf, which lets Kirby fire leaves. By shaking the Wii Remote whilst inhaling, Kirby can perform a Super Inhale, which allows him to swallow larger enemies and blocks or multiple small ones. The game introduces more powerful, though temporary copy abilities called "Super Abilities," which are able to destroy multiple enemies at once as well as parts of the environment. Throughout the game Kirby can collect food items which recover health and earn extra lives by collection 1UP items, or by collecting 100 stars. There are also various items Kirby can use, such as keys used to unlock hidden areas, a cannon which automatically fires, a trumpet that produces a protective barrier overhead and a large boot that lets Kirby hop across enemies and spiked areas (similar to the Kuribo's Shoe in *Super Mario Bros. 3*). At the end of each stage, a bonus game is played in which players time a button press to jump as high as possible to earn additional items.*[11]

The game features drop-in cooperative multiplayer for up to three individual players. These players can either play as a differently colored Kirby or as one of three unique characters; Meta Knight, King Dedede, and Waddle Dee. Each of these characters possess traits from multiple Copy Abilities, though only Kirbys are able to inhale enemies and use Copy Abilities. Players are able to ride on top of each other and share health items. Unlike most multiplayer games, all players share from a pool of lives, with each player requiring one life to enter the game. If one of the three additional players loses all of their health, they can spend an extra life to rejoin the game. However, if player one dies, gameplay resumes at the last checkpoint.*[10]

Throughout the game, players can seek out Energy Spheres used to power up the Lor Starcutter, which serves as the game's hubworld. Finding these spheres often requires keen exploration, wits or skill, or the use of a specific ability or item. There are also challenging warp areas that are found whilst using Super Abilities, requiring players to escape a fast approaching black void before facing off against a miniboss to earn additional Energy Spheres. Collecting enough of these spheres unlock Challenge Rooms, which test Kirby's mastery of a specific ability, Copy Ability Rooms, which let Kirby choose Copy Abilities, and two minigames; Ninja Dojo and Scope Shot. Ninja Dojo tasks players with shaking the Wii Remote to fire ninja stars at targets whilst Scope Shot tasks players with defeating a large robot within a time limit. Clearing the game unlocks Extra Mode, a more challenging remix of the main game, and The Arena, which sees players fighting all the bosses in a random order. Clearing the Extra Mode unlocks The True Arena, where players fight the harder versions of bosses, along with three extra boss characters.

2.27.2 Plot

The story begins with Kirby carrying a cake, while King Dedede and Waddle Dee chase him. They run past Meta Knight, who is reading a book under a tree. Kirby and the gang sees a ship flying out of a wormhole, and they go to investigate. They enter the ship and encounter Magolor, a creature who discovers that the five vital pieces of his ship, the Lor Starcutter, along with 120 energy spheres, have been scattered across the planet. With Magolor offering them a trip to his homeworld of Halcandra should they help fix his ship, Kirby and his friends set off to recover the lost pieces of his ship. After retrieving the main pieces, they travel to Halcandra, where they are attacked by a four headed dragon named Landia. Magolor claims Landia is an evil beast that has taken over Halcandra and sends Kirby to defeat it. However, after Landia is defeated, Magolor reveals his true motive was to steal the Master Crown on its head and become all powerful, with intent of making the entire universe bow before him, beginning with Popstar. Teaming up with Landia, who is split into four dragons, Kirby and his friends confront Magolor in a final battle and manage to destroy the Master Crown, taking Magolor with it (though the Kirby's 20th anniversary collection reveals that he survived and has reformed). With peace restored to the universe, Kirby and friends are returned to Popstar whilst the Landia dragons take the Lor Starcutter and return home.

2.27.3 Development

Development on a new *Kirby* title began immediately after the release of the 2000 game *Kirby 64: The Crystal Shards* as a title for the Nintendo 64. The game underwent an 11-year development period in which three different proposed versions of the game were developed and then scrapped.*[12] The first build was similar to the graphical and gameplay style of *Kirby 64*, rendered in 3D but using traditional 2D side-scrolling gameplay. The game would also support multiplayer with up to four players.*[9] This build was demonstrated at E3 in 2005, and was set for release later that year. However, difficulty with programming four-player led to this version being scrapped. The second build placed Kirby in a 3D environment with open world-style gameplay, and the third build returned to side-scrolling gameplay but had the graphical style of a pop-up book. The development team realized that the failure of the first three attempts were caused by too much focus on multiplayer, so focus was shifted almost exclusively to the single-player experience.*[12] Development of the final version accelerated in October 2010, when the game began to take form.*[12]

Kirby 's long development caused the game to frequently appear and then disappear from upcoming game lists. On September 14, 2006, the *Kirby* game appeared on a list of upcoming Wii games, named *Hoshi no Kābī* (星のカービィ lit. "*Kirby of the Stars*"), set for release in Japan. The December 2006 issue of *Nintendo Power* removed *Kirby* from its list of GameCube releases, but did not place it on its list of Wii releases. Matt Casamassina of IGN, posting on his blog, furthered the idea of a Wii release by stating that it would indeed be released for the Wii in 2007.*[13] He compared it to *Donkey Kong Barrel Blast*, another game that was originally announced as a GameCube title, but eventually released on Wii. While the game did not appear at E3 2007, Beth Llewelyn of Nintendo of America confirmed the game "had not been abandoned." *[14] The December 2007 issue of *Official Nintendo Magazine* claimed that a *Kirby* game for Wii was not in development.*[15] On May 7, 2010, Nintendo confirmed that a *Kirby* Wii title was still in the making.*[16]*[17]

Nintendo announced the game *Kirby's Epic Yarn* at E3 2010, a completely separate title that was in development by Good-Feel. The 2005 *Kirby* game was then presumed to have been canceled until a Financial Results Briefing on January 28, 2011 re-announced the game with a release date set within the same year.*[18] At E3 2011, the game was demonstrated in playable form under the tentative title *Kirby Wii*.*[19] The game was later renamed *Kirby's Return to Dream Land* in North America*[20] and *Kirby's Adventure Wii* in Europe,*[21] and *Hoshi no Kirby Wii* in Japan. The music is composed by Jun Ishikawa and Hirokazu Ando with an official soundtrack called *Kirby Wii Music Selection* featuring 45 musical pieces from the game.

2.27.4   Reception

Kirby's Return to Dream Land has received mostly positive reviews, with many praising the game's cooperative gameplay, graphics, and sounds but criticizing its lack of difficulty. Destructoid awarded the game a perfect score of 10/10, claiming, "Videogames simply do not get any more pure than Kirby's Return to Dream Land." *[25] Game Informer gave it an 8.5 out of 10, noting that, "While it doesn't have the challenge of Donkey Kong Country Returns or the charming art style of Kirby Epic Yarn, Kirby's Return To Dream Land is another formidable entry in a line

of great side-scrolling Wii platformers." *[26] IGN was a bit less positive, giving it a score of 7.5 out of 10, criticizing the game's lack of difficulty while stating it fits perfectly for a younger audience.*[28]

2.27.5   References

[1]   "Kirby's Return to Dreamland". Nintendo of America. Retrieved 23 August 2011.

[2]   発売カレンダー[Release Calendar] (in Japanese). Nintendo. Retrieved 1 October 2011.

[3]   "Super Kirby goes multiplayer in Kirby's Adventure Wii". Nintendo of Europe. 17 August 2011. Retrieved 20 August 2011.

[4]   "Wii Games - Kirby's Adventure Wii - Nintendo.com.au". Nintendo Australia. Retrieved 30 November 2011.

[5]   "Wii ディスクソフト（ダウンロ ド版について）". *Nintendo* (in Japanese). Nintendo. Retrieved 15 January 2015.

[6]   Whitehead, Thomas (2015-02-16). "Nintendo Download: 19th February (Europe)". Nintendo Life. Retrieved 2015-02-16.

[7]   Wassenaar, Troy (February 19, 2015). "AUSTRALIAN NINTENDO DOWNLOAD UPDATES (19/2) – DISCOUNT WAREHOUSE". *Vooks*. Retrieved 11 April 2015.

[8]   "Kirby Wii". E3 Nintendo. Retrieved 20 August 2011.

[9]   "GameCube Games of 2005". IGN. 2005-01-31. Retrieved 2007-07-13.

[10]   Slate, Chris (September 2011). "Fearsome Foursome". *Nintendo Power* (Nintendo) **271**: 42–47.

[11]   Claibron, Samuel (7 June 2011). "E3 2011: Kirby Wii -- Same Ol' Kirby". Retrieved 22 August 2011.

[12]   "Iwata Asks: Kirby's Return to Dream Land". Nintendo of America. Retrieved 25 October 2011.

[13]   "Having a Blast on Wii". February 2, 2007. Retrieved 2007-07-13.

[14]   "Disaster coming along, Project H.A.M.M.E.R. on hold, Smash online not confirmed, and Kirby!". *GoNintendo*. July 12, 2007. Retrieved July 13, 2007.

[15]   "Say it ain't so - No Kirby game in development for Wii". *GoNintendo*. December 17, 2007. Retrieved December 17, 2007.

[16]   "Financial Results Briefing for the Three-Month Period ended June 2009" (PDF). *NCL*. July 31, 2009. Retrieved August 4, 2009.

[17] "Kirby Wii Still in Development, But Release Still TBA". 1up.com. 2010-05-07. Retrieved 2010-06-28.

[18] "Third Quarter Financial Results Briefing for Fiscal Year Ending March 2011". Nintendo of Japan. 28 January 2011. Retrieved 20 August 2011.

[19] Tom Mc Shea. "E3 2011: Kirby Wii Preview Hands-On". GameSpot.

[20] Spencer (21 July 2011). "Kirby Wii's actual name is...". Siliconera.

[21] "Nintendo announces packed 2011 line-up of upcoming games". Nintendo. 17 August 2011.

[22] "Kirby's Return to Dream Land for Wii". *GameRankings*. CBC Interactive. Retrieved April 21, 2014.

[23] "Kirby's Return to Dream Land for Wii Reviews". *Metacritic*. CBC Interactive. Retrieved April 21, 2014.

[24] Parish, Jeremy (October 24, 2011). "Review: Kirby Treads Familiar Ground in His Return to Dreamland". 1UP.com. Retrieved April 21, 2014.

[25] Sterling, Jim (October 23, 2011). "Review: Kirby's Return to Dream Land". Destructoid. Retrieved April 21, 2014.

[26] Ryckert, Dan (October 24, 2011). "Classic Gameplay Meets Great Co-Op - Kirby's Return to Dream Land". Game Informer. Retrieved April 21, 2014.

[27] "Kirby's Return to Dream Land Review". GameTrailers. October 25, 2011. Retrieved April 21, 2014.

[28] George, Richard (October 24, 2011). "Kirby's Return to Dream Land Review". *IGN*. IGN Entertainment. Retrieved April 21, 2014.

2.27.6 External links

- *Kirby's Return to Dream Land* Official Website

- *Kirby's Adventure Wii* Official Website

- *Hoshi no Kirby Wii* Official Website

2.28 Mercs Inc

Mercs Inc was a cancelled video game developed by Danger Close Games. It was supposed to be the sequel to 2005's *Mercenaries: Playground of Destruction* and 2008's *Mercenaries 2: World in Flames*. It is a multiplayer third-person shooter.[*][1]

2.28.1 Development

The game was first rumored to be in development by Pandemic Studios as either *Project X* or *Y* which were rumored to be based on the *Mercenaries* franchise. Footage of the game was first leaked onto the internet on November 24, 2009, the footage was a multiplayer trailer of the game.[*][2] The game was then officially announced by Electronic Arts less than a day later though without a proper platform or release date.[*][3] The announcement of the game came less than a week after the closure of Pandemic Studios by Electronic Arts as it had been in development there under the name *Project Y*. The game was being developed by remaining Pandemic Studio staff members located at EA Los Angeles until the company was closed down in 2013.

2.28.2 References

[1] Pandemic Studios (November 24, 2009). "Pandemic Studios Announces 'Mercs Inc'". Electronic Arts. Retrieved November 21, 2009.

[2] Brian Ashcraft (November 24, 2009). "This Appears To Be A New Mercenaries Game". Kotaku. Retrieved November 21, 2009.

[3] Stephen Totilo (November 24, 2009). "EA Makes Mercs Inc, A New "Pandemic" Game, Official". Kotaku. Retrieved November 21, 2009.

2.29 Mya (program)

Mya was an intelligent personal assistant under development by Motorola. Proposed features for the program included the ability to read emails and answer questions 24 hours a day. Mya was intended to work with an internet service Motorola was developing called Myosphere, and was planned to be a paid service that would eventually be used by other mobile carriers. A female computer-generated character was created to represent Mya in advertising. While the quality of the character's animation was praised, it received criticism for being over sexualised.

Both the character and the program were announced to the public via an advertisement in March 2000, though the program was not ready for use at that time. Despite the announcement generating a considerable amount of attention, little was heard regarding the project in subsequent months. The program was never officially released nor cancelled, though the trademarks for both Myosphere and Mya were abandoned by Motorola in 2002. The name Mya was believed to be a play on the words 'My assistant'.

2.29.1 Proposed features and development

The Internet service that Mya was developed for was called Myosphere.[*][1][*][2] Motorola began development of Myosphere in 1998, and it had been described as a speech enabled service "which enables consumers to manage and control wireless and wireline communications from a single point of access using natural voice commands." [*][3] Several other companies had already announced plans for similar software at the time; Alan Reiter from *Wireless Internet and Mobile Computing* was puzzled at Motorola's announcement of Myosphere, saying "They're kind of late to the [voice activation] party. But the party is likely to be very big. ... Motorola's entry will help further legitimize the value of voice response systems. But it's a tough market, and it will take time." The term myosphere was "a play on the theme of connecting the elements of an individual's world, or sphere." [*][4]

Intended to provide a human-like interface to the Internet,[*][5][*][6] Mya was to be accessed via a toll-free telephone number and a pin code.[*][7] The program was designed to work with any phone, including landlines, but primarily for mobiles,[*][8] and was to be accessible 24 hours a day.[*][9] Mya was said to be able to answer questions on topics such as stock prices, news, sports, weather conditions, traffic, airline reservations, addresses, and appointments,[*][10] as well as being able to call contacts in a mobile phone's address book.[*][11]

Intended to be a paid service[*][12] that would be ready by December 2000, Motorola hoped that Mya would also eventually be used on Palm Pilot and by other mobile carriers.[*][9] In July 2000 Motorola was reported to be planning to work with Nuance Communications to internationalize Mya,[*][13] and that same month BellSouth was declared to be the first carrier to buy the service.[*][7] According to an article in *Popular Science* in August 2000, Motorola was spending "millions of dollars" on both the Mya character and the program.[*][14] Mya was originally programmed only for English,[*][9] though by April 2001 the program was being developed in six languages, and additionally Nippon Telegraph and Telephone were said to be working with Motorola to develop a Japanese version.[*][15][*][16] Mya was voiced by actress Gabrielle Carteris,[*][9][*][17] and mechanically altered to sound more digital.[*][18]

2.29.2 Character

To create a commercial for Mya, Motorola hired the McCann Erickson company, who in turn hired Digital Domain to create the character.[*][19] The design was described as a "big-budget" production,[*][1] though Digital Domain were only given three months to complete the project.[*][19]

Mya's physical representation: a tall, thin,[*][7] blonde, blue-eyed white female,[*][5][*][20] was created in the likeness of a human model,[*][21][*][22] Michelle Holgate.[*][23][*][1] The initial inspiration for Mya came from vintage pin-up girls.[*][19] The first representation of Mya had a very small waist and large breasts, and was said to resemble Jessica Rabbit, which did not impress either Motorola or McCann Erickson.[*][19] Motorola asked Digital Domain to make Mya look as human as possible yet still be obviously artificial. The first completed iteration of Mya was so realistic that Motorola asked for her to be made more obviously digital.[*][23] Viewers were reportedly not impressed with Mya because they thought she was a real person.[*][1][*][19][*][24] Digital Domain visual effects supervisor and animation director, Fred Raimondi, decided to remodel Mya's appearance to be "just to the left of real".

> You know how when you first looked at Max Headroom and were like 'What is that?' That's [the effect] we were after.
> Fred Raimondi[*][19]

Mya's hair was changed from brunette to bleach blonde.[*][19] Her short spiky hair style was said to resemble that of Serena Altschul.[*][17] According to Digital Domain, giving Mya hair that was longer than ear-length was not possible in the time they had been given, due to the difficulties of creating digital photorealistic hair.[*][19] Mya's final body shape was an almost exact copy of the original model's measurements.[*][19] Mya was typically seen wearing a silver pantsuit but also appeared in halter tops in some shots and dressed in an evening gown for her debut.[*][17] While Digital Domain staff wanted Mya to appear in a knee-length skirt with high boots, Motorola and McCann chose the pantsuit, due to its contemporary look.[*][19]

Digital Domain chose not to use motion capture for Mya's movements as they believed it would constrain the character too much. Instead they used rotoscoping to place their digital character over the real model.[*][23] The evening gown Motorola selected for Mya to debut in was described by Digital Domain as the most difficult item of clothing they could have chosen, due to its transparency and layering. To render 150 frames (equating to 5–6 seconds of actual footage) of Mya moving in the gown in low-resolution required approximately 6 hours of processing time; the final high-resolution shots took longer.[*][23] Mya's rendering was so complex she crashed the computers at Digital Domain several times.[*][23] Mya's creators said they had difficulty making Mya appear as if she were "alive", and focused intensely on movements, specular highlights and eye blinks in order to "bring her to life".[*][18] The specular highlights also had the intended effect of making Mya shine in an inhuman manner;[*][20] when the light hit Mya at certain angles, a rainbow would appear.[*][19] Mya's skin was described as "part china doll, part disco ball." [*][19] Her distinct shine was based on that of a china plate[*][1][*][24] that the commercial's director, Alex Proyas, had bought in Australia.[*][19] In some shots of Mya, images were deliberately downgraded and had scan lines added to make the character appear more artificial.[*][23]

Mya's visual representation, however, appeared solely during advertising and on her website. Only her voice was to be heard when using the actual program.[*][18] Demonstrations of Mya's abilities and images of the character could be viewed at the now defunct website, mya.com.[*][17][*][18] Raimondi said he believed the name Mya was a play on the words '**My a**ssistant',[*][23] as did Sidney Matrix in the book *Cyberpop: Digital Lifestyles and Commodity Culture.*[*][25]

2.29.3 Debut and appearances

Mya as she appears in a commercial

Mya made her debut on March 26, 2000 in an 60-second[*][18] advertisement shown during the 72nd Academy Awards.[*][9][*][18] The ad featured Mya dressed in her evening gown and wearing a headset.[*][17] In the ad Mya steps out of a stretch limousine[*][9] and walks down the red carpet for the show. The ad declared Mya to be "the darling of the e-world, the 24-hour talking Internet" [*][26] and stated that Mya's abilities would change users' lives.[*][10] Despite making her first advertising appearance in March, the Mya program was not scheduled to be ready until December 2000.[*][2][*][10][*][7]

Mya subsequently received considerable media attention,[*][18] and was featured on the front covers of *USA Today*, *InStyle*, *Wired*[*][2][*][19] and *Adbusters.*[*][27] One promotion for Mya showed Hugh Hefner sitting in a limousine with two Playboy Bunnies, asking Mya to read him his emails.[*][2][*][26]

The Mya program was on display at the Motorola wireless booth at COMDEX in April 2000, which was visited by then president Bill Clinton. Mya "chok[ed] up halfway" through a demonstration for the president and had to be restarted.[*][28]

In 2006 Sidney Matrix stated Mya "disappeared" after her debut commercial;[*][27] in August 2000, a Yankee Group vice-president stated the debut advertisement for Mya was a "great ad, but where have [Motorola] gone with it? ... The spot drove viewers to its website to demo the product ... but failed to market Mya further." [*][29] Mya was never documented to have been released, nor was there an announcement of the program's cancellation; Motorola abandoned their trademark for 'Mya' on September 19, 2002[*][30] and their trademark for 'Myosphere' on December 1, 2002.[*][31]

2.29.4 Reception

Character

The reception of the character was mixed. Libby Callaway from *New York Post* stated Mya was one of their favourite "virtual babes", and said she threatened to take Lara Croft's title as the internet's most popular pin-up girl, also describing Mya as "the world's first 'cyber assistant'".[*][17] Whilst admitting that the character of Mya was visually appealing, John Sullivan of *Wireless Insider* also stated that Motorola "went overboard" by trying to give the Mya program a character in the hopes she would become a celebrity in her own right, and accused Motorola of trying to mimic the success of Lara Croft.[*][8] Mya was described in the 2003 book *Data Made Flesh: Embodying Information* as "by far" the best-rendered and most self-assured digital woman.[*][18] Noah Robischon from *Entertainment Weekly* called her debut the second creepiest moment at that year's Academy Awards (the first being Angelina Jolie kissing her brother).[*][32][*][33]

Writing in *Popular Mechanics*, Tobey Grumet described Mya as a male-chauvinistic creation,[*][9] and she was cited in the 2006 book *Physical Culture, Power, and the Body* as an example of simulated sexualised females.[*][5] Sidney Matrix stated that Mya's seductive appearance and sultry voice "depended on, borrowed from and retrench[ed] sexist stereotypes" , and accused Motorola of normalising the assumption that technology users are both male and heterosexual.[*][20] Motorola's marketing director Julie Roth defended the design of Mya's appearance and voice, attributing it to market research of what would appeal to users.[*][9]

Mya's character was often compared to the female computer-simulated character for

Ananova,[*][21][*][17][*][18][*][5][*][2] a web-based news service that was being developed around the same time.[*][34]

Announced program

Though Mya's character was generally regarded as impressive, the underlying technology was described by Peggy Albright in *Wireless Week* as not surprising; Albright said Motorola was "latest company in recent weeks to introduce a voice-activated virtual assistant", as Mya was announced shortly after Microsoft had announced their MiPad, and Lucent had launched their Mobile Voice Activated Dialing software.[*][26] However, Tobey Grummet spoke highly of the program in anticipation of its release,[*][9] and Mya was described by Elliot Drucker of *Wireless Week* as a solution to the limitations of accessing the Internet on a mobile phone, without a keyboard or large colour display.[*][35] While regarding the program with interest, John Sullivan doubted that Mya would persuade people who were not already on the Internet to start using it, and stated that if Mya could only read emails and not actually converse with him, he would rather just read his emails himself.[*][8] Dawn Chmielewski from the Orange County Register called Mya a "crude interpretation of things to come", noting that speech technology at the time was not without its limitations.[*][36]

> You need to speak like a BBC broadcaster to be understood and use the vocabulary of a toddler to get what you want.
> Dawn Chmielewski[*][36]

The program won the "Most Innovative Telephony Application" award at the 20th Annual AVIOS Conference in April 2001.[*][37]

2.29.5   References

[1] Dixon, Steve (2007). *Digital Performance: A History of New Media in Theater, Dance, Performance Art, and Installation*. MIT Press. p. 263. ISBN 978-0-262-04235-2.

[2] Barboza, David (April 25, 2000). "Motorola hopes a computer-generated character will link the real world with the virtual one". *The New York Times*. Retrieved July 6, 2014.

[3] "Motorola trials Myosphere for Internet mobile comms". *v3.co.uk*. August 4, 1998. Archived from the original on 14 July 2014. Retrieved July 7, 2014.

[4] Zajac, Andrew (August 4, 1998). "Motorola Tests Software For Dialing Using Voice". *Chicago Tribune*. Retrieved July 10, 2015.

[5] Vertinsky, Patricia; Hargreaves, Jennifer (2006). *Physical Culture, Power, and the Body*. Routledge. p. 238-239. ISBN 978-0-415-36352-5.

[6] Michitaka, Hirose (2001). *Human-Computer Interaction – Interact'01*. IOS Press. p. 199. ISBN 978-1-58603-188-6.

[7] Grumet, Tobey (1 July 2000). "Cool toy - Mya desires your attention.". *Marketing magazine*. Retrieved 16 November 2014.

[8] Sullivan, John (April 17, 2000). "Motorola's Hot New Wireless Data Interface? Voice.". *Wireless Insider* **18** (16): 1.

[9] Grumet, Tobey (November 2000). "Digital Dame". *Popular Mechanics*: 37.

[10] ""Virtual Sensation," Mya Makes Her Debut At the Oscars; Motorola Announces Its First Cyber-Assistant". *Business Wire*. March 27, 2000. Retrieved June 23, 2014.

[11] Johnston, Margret (April 19, 2001). "Motorola's Wireless Net App Integrates Voice". *Computerworld*. Retrieved July 6, 2014.

[12] "CNN Transcript: How to Protect Your Privacy on the Net; What's at Stake in the Microsoft Antitrust Battle?". *CNN*. April 1, 2000. Retrieved June 26, 2014.

[13] "Motorola and Nuance team up for mobile voice Web solutions". *Intelligent Systems Report* (Lionheart Publishing, Inc.) **17** (7): 5. July 2000.

[14] Quain, John R (August 2000). "Anchors Away". *Popular Science*: 47. Retrieved 16 November 2014.

[15] "Motorola Japan and NTT Software Sign MOU to Design and Deliver VoiceXML Applications in Japan; Collaboration to Provide Japanese Market with Voice Access of the Internet". *Business Wire*. 23 April 2001. Retrieved 29 August 2014.

[16] "Motorola Strikes Internet Voice Partnership with NTT". *Japanese Telecom* **7** (4): 12. April 2001. Retrieved 29 August 2014.

[17] Callaway, Libby (June 4, 2000). "A Wave of Cyber Babes: Digital Models Spin the Web With Their Style Savvy". *New York Post*. Retrieved June 23, 2014.

[18] Mitchell, Robert; Thurtle, Phillip (2003). *Data Made Flesh: Embodying Information*. Routledge. ISBN 978-0-415-96905-5.

[19] Larsen, Elizabeth (October 2000). "Mya way". *I.D.* (F+W) **47** (6): 88–92.

[20] Matrix 2006, p. 111.

[21] Matrix 2006, p. 112.

[22] Robinson, Laura; Halle, David (2002). "Digitization, the Internet, and the Arts" (PDF). *Qualitative Sociology* **25** (3): 374. Retrieved 26 June 2014.

[23] Bizony, Piers (2001). *Digital Domain: The Leading Edge of Visual Effects*. Aurum Press. p. 137. ISBN 978-0-8230-7928-5.

[24] O'Mahony, Marie (2002). *Cyborg: The Man-Machine*. London: Thames & Hudson. p. 68. ISBN 978-0-500-28381-3.

[25] Matrix 2006, p. 110.

[26] Albright, Peggy (April 3, 2000). "Mya debuts at the Oscars" . *Wireless Week* **6** (14): 38.

[27] Matrix 2006, p. 109.

[28] McCullagh, Declan (April 19, 2000). "Clinton does Comdex" . *Wired*. Retrieved July 6, 2014.

[29] Elkin, Tobi (7 August 2000). "Motorola out to build brand image" . *Crain's Chicago Business* **23** (33). p. 20.

[30] "Mya" . *United States Patent and Trademark Office*. Retrieved July 7, 2014.

[31] "Myosphere" . *United States Patent and Trademark Office*. Retrieved July 7, 2014.

[32] Robischon, Noah (April 21, 2000). "O Sole Mya" . *Entertainment Weekly*. Retrieved June 23, 2014.

[33] Tyrrell, Rebecca (February 22, 2015). "The most awkward Oscar moments ever, from Liza Minnelli' s selfie snub to Angelina Jolie' s 'incest kiss' ". *Metro*. Retrieved July 10, 2015.

[34] Silver, David; Massanari, Adrienne (1 September 2006). *Critical Cyberculture Studies*. New York: New York University Press. p. 264. ISBN 978-0814740248.

[35] Drucker, Elliot (May 1, 2000). "Tackling the human interface" . *Wireless Week* **6** (18): 50.

[36] Chmielewski, Dawn (July 4, 2000). "TALKING HEADS Your computer is becoming just another pretty face" . *Orange County Register*.

[37] "The 20th Annual AVIOS Conference Announces Best of Show Award Winners" . *Business Wire*. April 24, 2001. Retrieved July 6, 2014.

Bibliography

- Matrix, Sidney (2006). *Cyberpop: Digital Lifestyles and Commodity Culture*. Routledge. ASIN B00CX0JFL0.

2.29.6 External links

- Official website at Internet Archive

2.30 Noctis

Noctis (Latin for "of night") is a computer space flight simulator featuring first-person visual exploration of an imaginary galaxy.

The player is manifested in *Noctis* as the pilot of spacecraft called a Stardrifter, capable of instantaneous interstellar travel. This allows travelling between stars, refuelling the Stardrifter from Lithium ion-ejecting stars, approaching planets in star systems and their moons, and even landing where it is physically possible. Many planets feature atmospheres and weather effects. Some harbour plants and animals, or even mysterious ruins. No goal or measure of success is imposed by the game: it simply allows to catalogue and annotate the player's discoveries in a common database of stellar bodies called the *GUIDE*, which is to be synchronised over the author's Internet server.

2.30.1 Gameplay

The *Noctis* galaxy, Feltyrion, is approximately 90 thousand light-years in radius, approximately double the radius of the Milky Way Galaxy. With the exception of landing on some world types (being gas giants, Substellar Objects, and unstable worlds), this galaxy is entirely open to be explored. Billions of worlds can be explored despite the program's relatively small size, due to its content being generated on the fly. The Noctis universe contains several types of planets and 12 types of stars, noted in-game with an S in front of the number of the star.

2.30.2 See also

- Procedural generation
- *No Man's Sky*
- *Frontier: Elite II*
- *Frontier: First Encounters*
- *Celestia*
- *Kerbal Space Program*
- *Space Engine*

2.30.3 References

[1] "Web Archive records of the earliest Noctis' homepage" . Retrieved 2009-04-30.

2.30.4 External links

- Official website

- *Noctis* at MobyGames

- Article on Noctis at *The Escapist*

2.31 Octopiler

Octopiler is IBM's prototype compiler to allow software developers to write code for Cell processors.

2.31.1 External links

- IBM Research Project - Compiler Technology for Scalable Architectures

- IBM Systems Journal - Using advanced compiler technology to exploit the performance of the Cell Broadband Engine architecture

- IBM's Octopiler, or, why the PS3 is running late on ArsTechnica

2.32 The Outsider (video game)

The Outsider is a "techno thriller" [4] video game in development for Microsoft Windows, PlayStation 3 and Xbox 360 that is set to feature storylines that change based on the player's actions. It was announced in 2005 as a David Braben game being developed by his company Frontier Developments. The release date had not officially been announced, [5] but during development the game was dropped by publisher Codemasters, leading to nearly 30 staff layoffs at the developer Frontier. After six years of development, in January 2011 David Braben confirmed that development had ceased, [3] but the game had not been cancelled. [6]

2.32.1 Gameplay

The Outsider is set in a crowded city based on present-day Washington, D.C. and its surrounding areas, including the CIA Headquarters in Langley, Virginia, Joint Base Andrews, and Newport News Shipbuilding. The player controls a CIA intelligence officer by the name of Jameson, who can use various hand-to-hand combat styles and weapons. The game's opening sequence depicts the character wrongly becoming a fugitive, and leaves the player to decide how to continue. [2]

2.32.2 Development

Frontier Developments has been focusing on new game-play elements that are now possible on the current generation of video game consoles. Several key proprietary technologies, which Frontier has been developing for some time, make their debut in *The Outsider* and bring the sort of freedom of action first seen in *Elite* up to date. [7]

The developers claim that the game abandons the traditional, prescriptive, mostly linear story of current generation games, and replaces it by simulating characters' motivations and aims. This gives the player genuine freedom to change the story outcomes. Frontier Developments says that each player will get a unique experience rather than simply switching between 'good' or 'evil'. [8] In an interview with the Gametheoryshow, Braben said that it is this level of freedom that will demonstrate what it is to be a "next-gen" video game. [9]

A new animation system is being developed for the game with the aim of giving a more realistic feel as it is adaptive and less scripted than typical animations. [5]

2.32.3 References

[1] "Interview with David Braben, Producer of LostWinds and Creator of the Elite Series". Diehard GameFAN. 11 June 2008. Retrieved 11 June 2008.

[2] "Frontier announces "The Outsider", its first 'next generation' project" (Press release). Frontier Developments. 16 September 2005. Retrieved 25 September 2007.

[3] "Exclusive: Frontier's The Outsider Cancelled?". Rock, Paper, Shotgun. 20 January 2011. Retrieved 1 September 2011.

[4] "Official details" (embedded swf). Frontier Developments. Retrieved 25 September 2007.

[5] Waters, Darren (9 August 2007). "What exactly is a next generation game?". BBC News. Retrieved 25 September 2007.

[6] "The Outsider 'not cancelled' | Game Development | News by Develop". Develop-online.net. Retrieved 1 September 2011.

[7] Stuart, Keith (16 September 2005). "Braben unveils Outsider". The Guardian.

[8] Houlihan, John (23 September 2005). "Xbox Interviews: David Braben - The Outsider". Computer and Video Games. Retrieved 25 September 2007.

[9] "Game Theory Episode 21 - Elite" (mp3). 21 August 2007. Retrieved 14 September 2007.

2.32.4 External links

- Official website

2.33 Prey 2

Prey 2 is a cancelled[*][2][*][3] first-person shooter video game published by Bethesda Softworks. It would have been the sequel to the 2006 video game *Prey*.[*][4]

2.33.1 Plot

The story of *Prey 2* was to focus on U.S. Marshal Killian Samuels, who starts the game on a passenger flight which suddenly crashes onto the Sphere (the crash is shown in the original *Prey*). At the end of a short battle with some aliens he is knocked unconscious, after which the plot jumps forward several years. Samuels is now a bounty hunter on the alien world Exodus. Though he is aware of his profession and has retained his skills, he has no memory of what happened in the time that passed since his abduction.[*][5] He initially believes himself to be the only human on Exodus until he runs into Domasi "Tommy" Tawodi (the protagonist of the original *Prey*), whom he has apparently met in the period he no longer remembers.[*][6] Killian then resumes his bounty hunter activities while recovering his memory.

2.33.2 Development

The sequel to 2006 game *Prey* was officially announced on March 14, 2011 as being developed by Human Head Studios using heavily modified Id Tech 4 engine.[*][7] "We are thrilled to be working with Bethesda on *Prey 2*", said Chris Rhinehart, project lead. "*Prey 2* will provide gamers the opportunity to explore a new facet of the Prey universe, one that offers fast-paced action in an open, alien world. We're excited to show gamers the title we have been working on and hope they will be as excited by this title as we are." [*][8]

Prey 2 was announced once before; shortly after the release of *Prey*, 3D Realms' Scott Miller confirmed that a sequel named *Prey 2* was already in development.[*][9] On March 17, 2008, Miller's brand-management group Radar Group was officially launched, along with the announcement that it is managing *Prey 2*, and that it slated for the PC, PlayStation 3 and Xbox 360. The rights were later transferred from Radar to Bethesda Softworks.[*][10] According to Pete Hines, the vice president of PR and marketing at Bethesda, the current version of the game is what the developers wanted to make and not what has been announced before by the Radar Group.[*][11]

On March 23, 2012, Dutch gaming website PS Focus reported a rumour that *Prey 2* has been cancelled by ZeniMax Media. When asked, Bethesda Softworks reported with no comment regarding the rumour on their Twitter page.[*][12][*][13][*][14][*][15][*][16] On April 19, 2012, Bethesda stated *Prey 2* will not be cancelled, but instead it will not make its scheduled 2012 release. This was because "the game's development has not progressed satisfactorily this past year and the game does not currently meet [their] quality standards.".[*][17] On August 20, 2012 the game was removed from the products page on Bethesda's website. A spokesperson from Bethesda informed Eurogamer that until they're ready to talk about the game more, the focus on the site is on their upcoming titles.[*][18]

In May 2013 Kotaku reported rumors that development has moved to Arkane Studios and that the development has been rebooted scrapping all of Human Head Studios work and with a targeted release of 2016. It has also been reported that Obsidian Entertainment worked on the game at one point for at least a few months.[*][19] In August 2013 Bethesda Softworks' Pete Hines denied rumors that Arkane Studios was working on the game.[*][20] However, on August 15, 2013, it was reported that leaked emails confirmed the game was in fact in development.[*][21][*][22][*][23][*][24]

On October 30, 2014 during PAX Australia, it was confirmed by Bethesda Softworks vice president Pete Hines that Prey 2 had subsequently been cancelled. Pete Hines statement as follows, "It was a game we believed in, but we never felt that it got to where it needed to be – we never saw a path to success if we finished it. It wasn't up to our quality standard, and we decided to cancel it. It's no longer in development. That wasn't an easy decision, but it's one that won't surprise many folks given that we hadn't been talking about it. Human Head Studios is no longer working on it. It's a franchise we still believe we can do something with we just need to see what that something is." [*][2][*][3] Tim Gerritsen, business development director at Human Head Studios, said, "While we are disappointed that we won't be able to deliver our vision of the game, we remain proud of our work on the franchise, which we feel speaks for itself, including the award-winning presentation of the game at E3 2011. We enjoyed working with the many talented people at Bethesda, and we wish them all the best of luck with any future plans they may have for the franchise." [*][25]

2.33.3 References

[1] Lapins, Vlad. "Prey 2 concept from Human Head Studio leaked the Internet". AngryGamer.ru (Russian)

[2] Murillo, Edwin (October 31, 2014). "Prey 2 has been officially cancelled, Bethesda confirms". *Gamer Headlines*.

Retrieved April 29, 2015.

[3] Healey, Nick (October 30, 2014). "Bethesda confirms Prey 2 cancelled". *CNET*. Retrieved April 29, 2015.

[4] Rosenberg, Adam (March 14, 2011). "'Prey 2' Bringing An 'Open, Alien World' For A 2012 Release From Bethesda Softworks". Multiplayerblog.mtv. Retrieved April 20, 2011.

[5] "Prey 2 Preview: Alien sequel". *Joystiq*. Retrieved June 4, 2011.

[6] "Prey 2 Story Introduction Interview HD". Retrieved June 4, 2011.

[7] D Deesing, Jonathan (April 18, 2011). "Prey 2 producer on taking new direction, with 'capable' id Tech 4". *Joystiq*. Engadget. Retrieved April 7, 2015.

[8] Sinicki, Joe (February 22, 1999). "Prey 2 will surprise you". Blastmagazine.com. Retrieved October 25, 2011.

[9] "Next-Gen People: Scott Miller". *next-gen.biz*. Edge. August 9, 2006. Retrieved June 3, 2009.

[10] Bethesda Announces Prey 2 for PC, PS3, 360 GameFront, March 14, 2011 (Article by Ron Whitaker)

[11] "Twitter / Pete Hines: It's worth clarifying that". *Twitter*. Retrieved October 25, 2011.

[12] "Jim Reilly - Status". *Twitter*. March 24, 2012. Retrieved April 7, 2015.

[13] Senior, Tom (March 26, 2012). "Prey 2 feared cancelled". *PC Gamer*. Retrieved April 7, 2015.

[14] Dutton, Fred (March 23, 2012). "Prey 2 cancelled – report • News •". Eurogamer.net. Retrieved April 9, 2012.

[15] "Prey 2 is mogelijk geannuleerd door Bethesda". PSFocus.nl. Retrieved April 9, 2012.

[16] "Bethesda won't deny Prey 2 cancellation rumours". Metro.co.uk. Retrieved April 9, 2012.

[17] Cullen, Johnny. "Bethesda: Prey 2 not canned, but it won't make 2012". VG247.

[18] Purchese, Robert. "Why Prey 2 was removed from Bethesda's website". Eurogamer.

[19] Schreier, Jason (May 31, 2013). "We Hear The People Behind Dishonored Are Now Working On Prey 2". *Kotaku*. Retrieved April 7, 2015.

[20] Grayson, Nathan (August 2, 2013). "Bethesda Talks Prey 2, Denies Arkane Involvement". *Rock, Paper, Shotgun*. Retrieved April 7, 2015.

[21] Schreier, Jason (August 15, 2013). "Leaked E-mails Suggest Bethesda Misled Gamers About Prey 2". *Kotaku*. Retrieved April 7, 2015.

[22] O'Brien, Lucy (August 15, 2013). "Report: Arkane is making Prey 2 after all". *IGN*. Retrieved April 7, 2015.

[23] Mallory, Jordan (August 15, 2013). "Report: Prey 2 reboot in development at Arkane Austin". *Joystiq*. Engadget. Retrieved April 7, 2015.

[24] Walker, John (August 15, 2015). "The Smell Of Bullshit: Arkane ARE Making Prey 2". *Rock, Paper, Shotgun*. Retrieved April 7, 2015.

[25] Martin, Matt (November 3, 2014). "Human Head "proud" of its work on cancelled Prey 2". *VG247*. Retrieved April 29, 2015.

2.33.4 External links

- Official site

2.34 Project Xanadu

"Xanadu Project" redirects here. For other uses, see Xanadu (disambiguation).

Project Xanadu was the first hypertext project, founded in 1960 by Ted Nelson. Administrators of Project Xanadu have declared it an improvement over the World Wide Web, with mission statement: "Today's popular software simulates paper. The World Wide Web (another imitation of paper) trivialises our original hypertext model with one-way ever-breaking links and no management of version or contents." [1]

Wired magazine published an article called "The Curse of Xanadu", calling Project Xanadu "the longest-running vaporware story in the history of the computer industry".[2] The first attempt at implementation began in 1960, but it was not until 1998 that an incomplete implementation was released. A version described as "a working deliverable", **OpenXanadu**, was made available in 2014.

2.34.1 History

During his first year as a graduate student at Harvard, Ted Nelson began implementing the system which contained the basic outline of what would become Project Xanadu: a word processor capable of storing multiple versions, and displaying the differences between these versions. Though he did not complete this implementation, a mockup of the system proved sufficient to inspire interest in others.

On top of this basic idea, Nelson wanted to facilitate nonsequential writing, in which the reader could choose his or her own path through an electronic document. He built upon this idea in a paper to the ACM in 1965, calling the new idea "zippered lists" . These zippered lists would allow compound documents to be formed from pieces of other documents, a concept named transclusion. In 1967, while working for Harcourt, Brace, he named his project Xanadu, in honour of the poem "Kubla Khan" by Samuel Taylor Coleridge.

1970s

Ted Nelson published his ideas in his 1974 book *Computer Lib/Dream Machines* and the 1981 *Literary Machines*.

Computer Lib/Dream Machines is written in a non-sequential fashion: it is a compilation of Nelson's thoughts about computing, among other topics, in no particular order. It contains two books, printed back to back, to be flipped between. *Computer Lib* contains Nelson's thoughts on topics which angered him, while *Dream Machines* discusses his hopes for the potential of computers to assist the arts.

In 1972, Cal Daniels completed the first demonstration version of the Xanadu software on a computer Nelson had rented for the purpose, though Nelson soon ran out of money. In 1974, with the advent of computer networking, Nelson refined his thoughts about Xanadu into a centralised source of information, calling it a "docuverse".

In the summer of 1979, Nelson led the latest group of his followers, Roger Gregory, Mark S. Miller and Stuart Greene, to Swarthmore, Pennsylvania. In a house rented by Greene, they hashed out their ideas for Xanadu; but at the end of the summer the group went their separate ways. Miller and Gregory created an addressing system based on transfinite numbers which they called tumblers, which allowed any part of a file to be referenced.

1980s

The group continued their work, almost to the point of bankruptcy. In 1983, however, Nelson met John Walker, founder of Autodesk, at The Hackers Conference a conference originally for the people mentioned in Steven Levy's *Hackers,* and the group started working on Xanadu with Autodesk's financial backing.

According to economist Robin Hanson, in 1990 the first known corporate prediction market was used at Xanadu. Employees and consultants used it for example to bet on the cold fusion controversy at the time.

While at Autodesk, the group, led by Gregory, completed a version of the software, written in the C programming language, though the software did not work the way they wanted. However, this version of Xanadu was successfully demonstrated at The Hackers Conference and generated considerable interest. Then a newer group of programmers, hired from Xerox PARC, used the problems with this software as justification to rewrite the software in Smalltalk. This effectively split the group into two factions, and the decision to rewrite put a deadline imposed by Autodesk out of the team's reach. In August 1992, Autodesk divested the Xanadu group, which became the Xanadu Operating Company, which struggled due to internal conflicts and lack of investment.

Charles S. Smith, the founder of a company called Memex (named after a hypertext system proposed by Vannevar Bush[*][3]), hired many of the Xanadu programmers and licensed the Xanadu technology, though Memex soon faced financial difficulties, and the then-unpaid programmers left, taking the computers with them (the programmers were eventually paid). At around this time, Tim Berners-Lee was developing the World Wide Web. When the Web began to see large growth that Xanadu did not, Nelson's team grew defensive in the supposed rivalry that was emerging, but that they were losing. The 1995 *Wired* Magazine article "The Curse of Xanadu," provoked a harsh rebuttal from Nelson, but contention largely faded as the Web dominated Xanadu.[*][4]

1990s

In 1998, Nelson released the source code to Xanadu as Project Udanax,[*][5] in the hope that the techniques and algorithms used could help to overturn some software patents.

2000s

In 2007, Project Xanadu released XanaduSpace 1.0.

2010s

A version described as "a working deliverable" , OpenXanadu, was made available on the WorldWide Web in 2014. It is called open because "you can see all the parts" , but as of June 2014 the site stated that it was "not yet open source" . On the site, the creators claim that Tim Berners-Lee stole their idea, and that the World Wide Web is a "bizarre structure created by arbitrary initiatives of varied people and it has a terrible programming language" and that Web security is a "complex maze". They go on to say that Hypertext is designed to be paper, and that the World Wide Web allows nothing more than dead links to other dead pages.[*][6]

2.34.2   Original 17 rules

1. Every Xanadu server is uniquely and securely identified.

2. Every Xanadu server can be operated independently or in a network.

3. Every user is uniquely and securely identified.

4. Every user can search, retrieve, create and store documents.

5. Every document can consist of any number of parts each of which may be of any data type.

6. Every document can contain links of any type including virtual copies ("transclusions") to any other document in the system accessible to its owner.

7. Links are visible and can be followed from all endpoints.

8. Permission to link to a document is explicitly granted by the act of publication.

9. Every document can contain a royalty mechanism at any desired degree of granularity to ensure payment on any portion accessed, including virtual copies ("transclusions") of all or part of the document.

10. Every document is uniquely and securely identified.

11. Every document can have secure access controls.

12. Every document can be rapidly searched, stored and retrieved without user knowledge of where it is physically stored.

13. Every document is automatically moved to physical storage appropriate to its frequency of access from any given location.

14. Every document is automatically stored redundantly to maintain availability even in case of a disaster.

15. Every Xanadu service provider can charge their users at any rate they choose for the storage, retrieval and publishing of documents.

16. Every transaction is secure and auditable only by the parties to that transaction.

17. The Xanadu client-server communication protocol is an openly published standard. Third-party software development and integration is encouraged.*[7]

2.34.3   See also

- Enfilade (Xanadu)

- Hypertext

- Hypermedia

- Memex

- ENQUIRE

- Interpedia

- American Information Exchange

- Tent (protocol)

- Chantment, a similar proposed information system running on Artificial intelligence

2.34.4   Footnotes

[1] Project homepage

[2] Gary Wolf (June 1995). "The Curse of Xanadu". *WIRED* **3** (6).

[3] "As We May Think" - The original article from the *Atlantic Monthly* archives

[4] Reagle, Joseph Michael (2010). *Good Faith Collaboration: The Culture of Wikipedia*. Cambridge, MA: MIT Press. ISBN 978-0-262-01447-2.

[5] "Udanax Green".

[6] Xanadu Web page Sample document: "Origins", by Moe Juste "takes a while to open because it's downloading a lot"

[7] Xanadu FAQ: What requirements do Xanadu systems aim to meet?, 2002-04-12 by Andrew Pam

2.34.5   References

- *The Magical Place of Literary Memory: Xanadu* in Screening the Past, July 2005 by Belinda Barnet

- *The Curse of Xanadu*, Wired feature on Nelson and Xanadu

  - Published comments on that Wired article, including one from Ted Nelson

  - *Errors in "The Curse of Xanadu"* by Theodor Holm Nelson, Project Xanadu

2.34.6 External links

- Official website

- Xanadu Australia – an active site

- "Xanadu Products Due Next Year," by Jeff Merron. BIX online news report from the West Coast Computer Faire, 1988

- Ted Nelson Possiplex Internet Archive book reading video

2.35 Protein-coated disc

Protein-Coated Disc (**PCD**) is a theoretical optical disc technology currently being developed by Professor Venkatesan Renugopalakrishnan, Children's Hospital, Harvard Medical School. PCD would greatly increase storage over Holographic Versatile Disc optical disc systems. It involves coating a normal DVD with a special light-sensitive protein made from a genetically altered microbe, which would in principle allow storage of up to 50 Terabytes on one disc. Working with the Japanese NEC Corporation, Renugopalakrishnan's team created a prototype device and estimated in July, 2006 that a USB disk would be commercialised in 12 months and a DVD in 18 to 24 months.*[1] *[2]

The technology uses the photosynthetic pigment bacteriorhodopsin created from bacteria.

2.35.1 Technology

The information in such discs would be highly dense, due to being stored in proteins that are only a few nanometres across. However, a method to address individual protein molecules to read and write information to and from them would have to be developed in order to achieve the theoretical 50 TB capacity. Practically, capacity would probably be limited by the size that addressing light can be focused to, so a DVD-sized disc might be able to hold ~50 GB, or perhaps ~240 GB if near-field optics were used.*[3]

2.35.2 See also

- DVD

- HD DVD

- Blu-ray Disc

- HVD

- 3D optical data storage

- Protein

2.35.3 References

[1] Richard Blanchard (July 11, 2006). "50 terabyte flash drive made of bug protein". GetUSB info (& ABC News Australia). Retrieved 17 Mar 2012.

[2] Tuan Nguyen (July 12, 2006). "New Research Promises 50TB on DVD-size Discs". DailyTech.

[3] Evan Blass (July 12, 2006). "Protein Coated Discs". Engadget.

2.36 Rainbow Storage

Rainbow Storage is a developing paper-based data storage technique first demonstrated by Indian student Sainul Abideen in November 2006.*[1] Abideen received his MCA from MES Engineering College in Kuttipuram in Kerala's Malappuram district.

Initial newspaper reports of the technology were debunked by multiple technical sources, although Abideen says those reports were based on a misunderstanding of the technology. The paper meant to demonstrate the capability of storing relatively large amounts of data (and not necessarily in the gigabyte range) using textures and diagrams.*[2]

The Rainbow data storage technology claims to use geometric shapes such as triangles, circles and squares of various colors to store a large amount of data on ordinary paper or plastic surfaces. This would provide several advantages over current forms of optical- or magnetic data storage like less environmental pollution due to the biodegradability of paper, low cost and high capacity. Data could be stored on "Rainbow Versatile Disk" (RVD) or plastic/paper cards of any form factor (like SIM cards).*[3]

The technique is, nonetheless, widely regarded as a scam (see Criticism).

2.36.1 Criticism

Following the wide media attention this news received, some of the claims have been debunked by various experts; however, Sainul Abideen says that the articles are all based on misunderstandings.*[4] *[5]

Printing at 1,200 dots per inch (DPI) leads to a theoretical maximum of 1,440,000 colored dots per square inch. If a scanner can reliably distinguish between 256 unique colors (thus encoding one byte per dot), the maximum possible storage is approximately 140 megabytes for a sheet of A4 paper–much lower when the necessary error correction is employed. If the scanner were able to accurately distinguish between 16,777,216 colors (24 bits, or 3 bytes per dot), the capacity would triple, but it still falls well below the media stories' claims of several hundred gigabytes.

Printing this quantity of unique colors would require specialized equipment to generate many spot colors. The process color model used by most printers provides only four colors, with additional colors simulated by a halftone pattern.

At least one of three things must be true for the claim to be valid:

- The paper must be printed and scanned at a much higher resolution than 1,200 DPI,

- the printer and scanner must be able to accurately produce and distinguish between an extraordinary number of distinct color values, or

- the compression scheme must be a revolutionary lossless compression algorithm.

The theory is: If Rainbow's "geometric" algorithm is to be encoded and decoded by a computer, it would equally viable to store the compressed data on a conventional disk rather than printing it to paper or other non-digital medium. Printing something as dots on a page rather than bits on a disk will not change the underlying compression ratio, so a lossless compression algorithm that could store 250 gigabytes within a few hundred megabytes of data would be revolutionary indeed. Likewise, data can be compressed with *any* algorithm and subsequently printed to paper as colored dots. The amount of data that can be reliably stored in this way is limited by the printer and scanner, as described above.

2.36.2   Demonstrations

Sainul Abdeen demonstrated his technology to the college and members of the Indian press in the MES College of Engineering computer lab, Kerala, and was able to compress 450 sheets plain text from foolscap paper into a 1 inch square. He also demonstrated a 45-second audio clip compressed using this technology on to an A4 sheet. Depending on the sampling frequency, bit depth, and audio compression (if any), a 45-second audio clip can consist of anywhere from a few kilobytes to a few megabytes of data. Abideen claimed that the technology could be extended to 250 gigabytes by using specific materials and devices.

This technology is based on two principles:

Principle I "Every color or color combinations can be converted into some values and from the values the colors or color combinations can be regenerated" .

Principle II "Every different color or color combinations will produce different values" .

2.36.3   References

[1]   "Data Can Now Be Stored on Paper" by M. A. Siraj, *ArabNews* (published November 18, 2006; accessed November 29, 2006)

[2]   Paper storage man misunderstood *The Inquirer* article, 12 December 2006 (retrieved 15 December 2006.

[3]   "Store 256GB on an A4 sheet" by Chris Mellor, Techworld (published November 24, 2006; accessed November 29, 2006)

[4]   IT Soup: Scam of Indian student developing technology to store 450 GB of data on a sheet of paper By ITSoup (published November 25, 2006; accessed November 25, 2006)

[5]   "Can you get 256GB on an A4 sheet? No way!" By Chris Mellor, Techworld (published November 24, 2006; accessed November 29, 2006)

2.36.4   Absolute Rainbow Dots

Absolute rainbow dots are used to detect errors caused by scratches, and whether any fading has occurred. Absolute rainbow dots are predefined dots carrying a unique value. These dots can be inserted in the rainbow picture in pre-specified areas. If fading occurs these dot values will change accordingly, and at the reproduction stage this can be checked and corrected. Absolute rainbow dots will be microscopically small so that they occupy very little space in the rainbow picture. These will be colored differently so that each dot will have its own fixed unique value.

2.36.5   External links

- Sainul Abideen's home page (dead)

- Deccan Herald's article on Rainbow Storage (dead)

- Article in DailyTech,

- IT Soup: Scam of Indian student developing technology to store 450 GB of data on a sheet of paper

- Article in The Register

- IDM: Paper Storage Claims A Hoax? (dead)

2.37 Sadness (video game)

Sadness was a survival horror video game in development by Nibris for the Wii console and was one of the earliest titles announced for the system.[*][5] While the game initially drew positive attention for its unique gameplay concepts, such as black-and-white graphics and emphasis on psychological horror over violence, *Sadness* became notorious when no evidence of a playable build was ever publicly released during the four years it spent in development.[*][6][*][7] It was revealed that *Sadness* had entered development hell due to problems with deadlines and relationships with external developers,[*][2] leading to its eventual cancellation by 2010,[*][3] along with the permanent closure of the company.[*][6]

2.37.1 Concept

Sadness was promoted as a unique and realistic survival horror game that would "surprise players," focusing on psychological horror rather than violence, containing "associations with narcolepsy, nyctophobia and paranoid schizophrenia." [*][8] More notable was the announcement that the game would sport black-and-white visuals stylized as gothic horror.[*][9]

Nibris promised that *Sadness* would provide "extremely innovative game play," [*][2] fully utilizing the motion sensing capabilities of both the Wii Remote and the Nunchuk. For example, it was suggested that players would use the Wii Remote to wield a torch and wave it to scare off rats; swinging the controller like a lasso in order to throw a rope over a wall;[*][8] or picking up items by reaching out with the Wii Remote and grabbing them.[*][1] *Sadness* was also planned to have open-ended interactivity between the player and the game's objects, being able to use any available item as a weapon. Suggestions included breaking a glass bottle and using the shards as a knife, or breaking the leg off a chair and using it as a club.[*][2] The game would also not utilize in-game menus (all game saves would be done in the background) nor a HUD in favor of greater immersion.[*][1]

Story

Set in pre-World War I Ukraine, *Sadness* was to follow the player character Maria Lengyel, a Victorian era aristocrat of Polish-Hungarian descent who has to protect her son Alexander after their train to Lviv derails in the countryside.[*][2] Alexander, who is struck blind by the accident, begins to exhibit strange behavior that progressively worsens.[*][10] The game's scenarios and enemies, such as those based on the werewolf and the likho, are inspired by Slavic mythology.[*][1] In order to "make the player feel that he is participating in events [and] not merely playing a game," the game was planned to feature a branching storyline, influenced by the player's actions[*][1] and concluding with one out of ten possible endings.[*][10]

2.37.2 History

Sadness was announced by Nibris on March 7, 2006 as a title for Nintendo's Revolution (before its final name "Wii" was announced),[*][5] later releasing a live action concept trailer which demonstrated potential Wii Remote control in gameplay.[*][11] Nibris partnered with Frontline Studios, who would be in charge of the game programming, and Digital Amigos, who would develop the game visuals,[*][1] and it was reported that the game would be released by the fourth quarter of 2007.[*][12]

From 2007 onward, Nibris drew criticism from various websites and blogs for the lack of playable demos, trailers, or even screenshots of the game in action.[*][13][*][14][*][15] Speculation that *Sadness* was vaporware intensified following a number of events, particularly the announcement that Frontline Studios was no longer working on the project (citing "artistic differences")[*][2][*][13][*][16] and the game's delay to 2009.[*][17] Several announcements that were made by Nibris were also never realized, such as a new trailer by late 2007[*][18] or an appearance at the 2008 Game Developer's Conference.[*][19] Destructoid chose to stop covering *Sadness* following the delay, doubting the game's existence and citing the lack of proof outside of concept art and "fancy talk." [*][20] Joystiq labeled *Sadness* as both a "comedy of errors" [*][21] and a "public embarrassment." [*][22] Fog Studios, Nibris' marketing partner, responded to the accusations on vaporware on at least two occasions, once in June 2008[*][23] and again in September 2009,[*][24] insisting that *Sadness* development was still underway, but was in need of a publisher.

In May 2009, N-Europe interviewed Adam Artur Antolski, an ex-Nibris employee and scriptwriter for *Sadness,* who revealed that the game's delays were caused by prolonged dispute over the game design, which led to constant failures to meet deadlines.[*][2] Antolski asserted that because no consensus could be made among the staff and with Frontline, the only progress made during the first year was completion of the script, concept design, and "only one 3D object - some minecart I believe." [*][2] When asked regarding the announcement that Nibris would be present at that year's Electronic Entertainment Expo,[*][25] Antolski stated that "Nibris always was better in promoting than making anything." [*][2] As such, Nibris did not appear at E3 2009, and the game silently missed the projected Autumn 2009 release date.

Nibris' official website, which had not been updated since

its announcement of appearing at GDC 2008, closed in February 2010.*[26] On April 5, N-Europe reported that Arkadiusz Reikowski, one of the game's music composers, released some of his unfinished work to the public*[27] and confirmed that *Sadness* had been abandoned by Nibris and was no longer in the works.*[3] In October Nibris itself transformed into a coordinator for the European Center of Games, ceasing game development permanently, and remaining staff and projects were also reported to have been handed over to Bloober Team, another game developer.*[6]

On May 20, 2014, *Nintendo Life* published an interview with indie developers Randy Freer and Jeremy Kleve, of HullBreach Studios and Cthulhi Games respectively, who claimed they had officially acquired the *Sadness* intellectual property from Nibris and were planning to collaboratively re-develop the game's concept from scratch for a 2016 Wii U-exclusive release. The version would have utilized the Unity game engine.*[28] The next day, Freer and Kleve retracted their claim, stating that they had failed to acquire the rights to the property.*[29]

2.37.3 References

[1] Seff, Micah (27 July 2006). "Our Sadness Grows". IGN. Retrieved 6 October 2010.

[2] Phillips, Tom (13 May 2009). "Nibris - The Full Interview". N-Europe. Archived from the original on 16 May 2009. Retrieved 5 October 2010.

[3] Phillips, Tom (5 April 2010). "Sadness Officially Cancelled". N-Europe.

[4] "GAMEBRYO BREATHES LIFE INTO Wii' S FIRST BLACK-AND-WHITE HORROR TITLE". Emergent. 14 April 2008.

[5] "New Sadness details". GoNintendo. 7 March 2006. Retrieved 10 February 2011.

[6] Whincup, Nathan (19 October 2010). "News: Nibris Bites The Bullet Bill". N-Europe. Retrieved 11 February 2011.

[7] Fletcher, JC (25 May 2010). "Composer says Wii's 'Sadness' was cancelled, releases soundtrack". Joystiq. Retrieved 12 February 2011.

[8] Casamassina, Matt (8 March 2006). "Upstart Studio Targets Revolution". IGN. Retrieved 6 October 2010.

[9] "Sadness In Black and White". Advanced Media Network.

[10] Ransom-Wiley, James (16 February 2007). "Mysterious Wii 'Sadness' explained!". Joystiq. Retrieved 12 February 2011.

[11] Sadness E3 Trailer at IGN. 10 May 2006.

[12] Phillips, Tom (30 July 2006). "News: Sadness Music & Details". N-Europe. Retrieved 12 February 2011.

[13] Whincup, Nathan (9 April 2007). "Rumour: Frontline Studios Drop Sadness?". N-Europe.

[14] "Sadness vaporware from the start?". Giant Bomb. 6 May 2009.

[15] Fletcher, JC (30 April 2007). "New Sadness artwork is not screenshots". Joystiq. Retrieved 12 February 2011.

[16] "Delayed Video Games". Giant Bomb. 6 March 2009. Retrieved 12 February 2011.

[17] Grant, Christopher (2 January 2008). "Sadness delayed again, vaporware suspicions refuse to go away". Joystiq. Retrieved 12 February 2011.

[18] Karabinus, Alisha (31 December 2007). "A look back through time: Sadness". Joystiq. Retrieved 12 February 2011.

[19] Phillips, Tom (5 May 2009). "E3 2009: Sadness Surfacing At E3?". N-Europe. Retrieved 12 February 2011.

[20] Chester, Nick (3 November 2007). "Nibris' Wii title, Sadness 'delayed' until 2009, probably doesn't exist". Destructoid.

[21] Greenhough, Chris (1 November 2007). "Sadness: a chronology of disappointment". Joystiq. Retrieved 12 February 2011.

[22] Sliwinski, Alexander (3 February 2010). "Sadness website is gone ... don't even feign shock". Joystiq. Retrieved 12 February 2011.

[23] "Sadness Is "Real"". Kotaku. 20 June 2008.

[24] "Sadness still alive...sort of". GoNintendo. 9 September 2009.

[25] Fletcher, JC (6 May 2009). "Prepare for Sadness: Nibris showing something at E3". Joystiq. Retrieved 12 February 2011.

[26] Sterling, Jim (February 3, 2010). "Sadness is bulls*t: Nibris' official website closes". Destroid. Retrieved February 2, 2011.

[27] "Posłuchaj muzyki z gry Sadness!". Gra muzyka. 3 April 2010.

[28] Whitehead, Thomas (2014-05-20). "Exclusive: Lost Wii Game, Sadness, To Be Revived as Wii U Exclusive". Nintendo Life. Retrieved 2014-05-21.

[29] Matulef, Jeffrey (21 May 2014). "Wii vaporware horror game Sadness is resurrected for Wii U". *Eurogamer*. Gamer Network. Retrieved 21 May 2014.

2.38   Samsung Ativ Q

Samsung Ativ Q

The **Samsung Ativ Q** was a 13.3-inch convertible laptop to be manufactured by Samsung. Unveiled at a *Samsung Premiere* event on June 20, 2013, the tablet was to run Windows 8, but also shipped with software that also allowed it to run the Android operating system. The Ativ Q's hardware was also distinguished by multiple folding states and a high resolution display.

Samsung announced that the Ativ Q would be released in the third quarter of 2013, with a representative indicating that it would be out in time for the back to school season. However, in August 2013, a South Korean news outlet reported that the release of the device would be indefinitely delayed due to patent issues relating to its Android emulation system.*[1]*[2] Samsung has not made any statements regarding the Ativ Q's release since.

2.38.1   Specifications

Hardware

The Ativ Q's design incorporates a unique, rugged hinge (which also houses the CPU) that can be used to tilt the screen into a number of different positions, such as flipping it over entirely to use it like a stand, having it "float" above the keyboard on an angle, or in a traditional laptop-styled position. Due to the lack of space, a pointing stick is offered instead of a trackpad.*[3]*[4]

The Ativ Q uses a 4th generation (Haswell), 2.6 GHz Intel Core i5 4200U processor with 4 GB of RAM. The device

features at 13.3-inch capacitive touchscreen with a resolution of 3200×1800 at 275 ppi, and will also ship with an S Pen stylus.*[1]

Software

While Samsung's presentation showcased the Ativ Q running Windows 8.1, demo units of the Ativ Q at its launch event ran Windows 8. The Ativ Q was to ship with a stock version of Android 4.2.2 running inside a virtual machine, accessible from within the Windows environment. Shortcuts and a keyboard button are provided for switching to the Android environment, files can be shared between the two environments, and Android apps can also be pinned to the Windows Start screen. The Ativ Q is also bundled with Samsung's "SideSync" software for linking to and controlling other Samsung smartphones and tablets with Android.*[3]*[5]

2.38.2   Release

Although Samsung initially announced a late-2013 release in time for the back to school season,*[1] the South Korean edition of *ZDNet* reported in August 2013 that the Ativ Q's release would be delayed or cancelled due to patent issues surrounding its dual-OS functionality.*[2]*[6]

In March 2014, it was reported that both Microsoft and Google were colluding to prevent the release of devices shipping with both Windows and Android, in order to protect their respective market shares and application ecosystems. Pressure from the companies had reportedly resulted in Asus discontinuing its line of similar Windows/Android dual-boot products, including the Transformer Book Duet, which was similarly left in a vaporware state.*[7]*[8]

2.38.3   See also

- Vaporware

2.38.4   References

[1] "Samsung ATIV Q: hands-on with the company's new Windows-Android slider (video)". *Engadget*. Retrieved 21 June 2013.

[2] "Has Samsung's Ativ Q Android/Windows 8 hybrid been KO'd by patent woes?". *TechRadar*. Retrieved 29 August 2013.

[3] "Samsung ATIV Q Hands On: 3200 x 1800 13.3" Tablet Running Windows 8 and Android". *Anandtech*. Retrieved 21 June 2013.

[4] "Samsung ATIV Q review - hands on with the Windows and Android tablet". *Expert Reviews*. Retrieved 21 June 2013.

[5] "Samsung Unveils Ativ Q, Ativ Tab 3 Windows 8 Tablets". *PC Magazine*. Retrieved 21 June 2013.

[6] "[단독] 삼성괴물노트북' 아티브 Q' 전면보류". *ZDNet Korea* (in Korean). Mega News. Retrieved 29 August 2013.

[7] "Microsoft and Google ruin Intel's plan for dual-OS tablets". *The Verge*. Retrieved 17 March 2014.

[8] "Google and Microsoft are out to stop dual-boot Windows/Android devices". *Ars Technica*. Retrieved 17 March 2014.

2.39 Scratch: The Ultimate DJ

Scratch: The Ultimate DJ was a music video game announced by Genius Products in 2008. Similarly to Konami's Beatmania series, it would have employed a specialized turntable controller (called the "Scratch Deck"), which would have allowed the player to follow along to the rhythm game while simulating common DJ techniques, such as scratching.[2][3]

The game had been held up in legal action between publisher Genius Products and former developer 7 Studios. A replacement developer, Bedlam Games had been announced following the lawsuit and the source code was returned to Genius as per a legal order.[4][5]

On May 26, 2010, Numark announced that the game was coming to the iPhone, iPad and Microsoft Windows and were due for release later that year.[1] However, nothing was released.

In September 2011, Bedlam Games laid off 90% of its staff and later shut its doors.[6]

Since the closure of Bedlam, none of the parties have given any sort of statement regarding the future of the game. As of June 11, 2014 the website for *Scratch* no longer exists.

2.39.1 Legal conflicts

On April 15, 2009, the publishers of *Scratch: The Ultimate DJ*, Genius Products and Numark, sued against *Scratch*'s developer, 7 Studios and Activision. The lawsuit contends that Activision purchased 7 Studios to both gain access to proprietary technology and to delay publication of the game so *DJ Hero* could come out first.[7] The Los Angeles Superior Court in which the suit was filed did not grant the requested restraining order against Activision on *DJ Hero*. Activision states that *Scratch* was already delayed by as early as October 2008, before they made contact with 7 Studios,

and their acquisition of the developers did not impede them from completing *Scratch*.[8] However, on April 20, the court reversed its decision, awarding Genius and Numark a temporary restraining order, and ordered the "immediate return" of all of the material from 7 Studios from Activision,[9] including all source code related to *Scratch*.[10] 7 Studios subsequently filed a counter-suit against Genius Products, claiming that they engaged in "unlawful and unsavoury business practices" that limited 7 Studios from completing the game as planned.[11]

2.39.2 Set list

The songs below had been confirmed to be in the game.[12][13][14]

In addition to these songs, Genius had announced that the game would include music from Gorillaz, Deltron 3030, Snoop Dogg and Mix Master Mike.[15]

2.39.3 See also

- *DJ Hero*, a similar game released that was developed by FreeStyleGames for Activision

2.39.4 References

[1] JC Fletcher (26 May 2010). "Scratch: The Ultimate DJ remixed for iPhone, iPad, and PC [update]".

[2] "Announcement of Scratch".

[3] "Yahoo announcement of Scratch".

[4] Tor Thorson (6 August 2009). "Scratch delayed until 2010, new dev on board".

[5] Mike Fahey (9 May 2009). "Judge Orders 7 Studios To Hand Over Scratch DJ Source Code".

[6] JC Fletcher (22 August 2011). "D&D Daggerdale studio Bedlam Games 'effectively shuttered'".

[7] "Genius Products, Numark Sue 7 Studios, Activision Over Scratch 'Withholding'". Gamasutra. 2009-04-15. Retrieved 2009-04-16.

[8] Alexander, Leigh (2009-04-16). "Activision: Genius Products Suit Just 'An Attempt To Place Blame' For Scratch Delay". Gamasutra. Retrieved 2009-04-16.

[9] "InPlay". Briefing.com. 2009-04-20. Retrieved 2009-04-20.

[10] Cifaldi, Frank (2009-05-11). "Judge orders remaining Scratch DJ code returned". GamesIndustry.biz. Retrieved 2009-05-11.

[11] Lee, James (2009-04-24). "7 Studios counter-sues Genius". GamesIndustry.biz. Retrieved 2009-04-24.

[12] http://consolehero.com/2009/05/11/scratch-mp3-functional-turntable-info/#more-1573

[13] http://consolehero.com/2009/05/21/dj-hero-scratch-songs-revealed/

[14] http://kotaku.com/5365820/scratch-the-ultimate-dj-track-list-grows-by-eight

[15] "Genius Drops First Tracks for Scratch: The Ultimate DJ".

2.39.5 External links

- Official "Scratch: The Ultimate DJ" website

- *Scratch: The Ultimate DJ* at IGN

2.40 The Shadow of Aten

The Shadow of Aten was an action-thriller video game from Silicon Garage Arts. The player was to take the role of a British ex-agent, Allan Scott, investigating the murder of a renowned British archaeologist in Cairo in the year 1936.

2.40.1 Cancellation

Silicon Garage Art's website has been disabled, and various gaming sites have labeled The Shadow of Aten as cancelled.

2.41 Shenmue Online

Shenmue Online (シェンム オンライン 「莎木OL」, Shenmū Onrain) was an announced MMORPG where players participate in scenarios from *Shenmue II*, joining one of three clans, led by Shen Hua, Xiu Ying, and Wu Ying Ren, all significant characters from the story.

2.41.1 Development history

In 2004, *Shenmue Online* was announced to be development in joint venture between Sega Japan and JC Entertainment of Korea.[1] A private beta testing period was due to begin in Korea in November 2005, with a public beta in China in spring 2005, before a release later in 2005.[2] When *Shenmue Online* was first announced, under the development of JC Entertainment, the screenshots released were considered by many *Shenmue* fans to be disappointing graphically, especially as being part of a game series that boasted groundbreaking graphics on the Dreamcast.

In 2005, JC Entertainment pulled out of the development of *Shenmue Online*, and as they owned 50% of the properties of *Shenmue Online*, there was legal debate over who had the rights to the game.[3] However, in November 2005, it was confirmed in an interview with Yu Suzuki that even though the Korean company JCE had announced their withdrawal from development, since Shenmue Online is a Sega venture, they will fulfil their duty of completing it. Yu says despite the internet rumours, development is well on its way and following their original goals and target. He adds that Sega Sammy will make an official announcement about its progress in the future.[4]

At the China Joy convention in July 2006 with the development under the guise of a Taiwan based company, a trailer was shown, which showcased features including mini games and combat with marked graphical improvements.[5]

In 2006, 6 new screens from the gameplay along with a 14 minute video were released to the public during the 4th China Expo. These shots showed further improvement on graphical issues, comparing the graphics to other MMOs and even home versions of *Shenmue*.

Cancellation

Since 2007, news about Shenmue Online has slowly declined and reports of its cancellation have appeared on Destructoid[6] and Wired News.[7] Sega have yet to publicly confirm or deny the reports.

2.41.2 Gameplay

Quoted from an interpretation of Yu Suzuki's 2005-11-05 interview:

2.41.3 Notes

[1] "Shenmue Online Officially Announced".

[2] "Update: Shenmue Online".

[3] "Who's got the rights to Shenmue Online?".

[4] "Yu Suzuki Commentary".

[5] "Shenmue Online resurfaces".

[6] "Shenmue Online canceled: Who will move all of these online crates now?".

[7] Arendt, Susan (2007-08-07). "Rumor:Shenmue Gone for Good". *Wired*.

2.41.4   External links

- Official *Shenmue Online* website

- Official *Shenmue* website

2.42   Six Days in Fallujah

For other uses, see SDIF (disambiguation).

Six Days in Fallujah (*SDIF*) is an unreleased historical third-person shooter video game developed and left unreleased by Atomic Games. Described by Atomic Games as a tactical shooter, it is the first video game to focus directly on the Iraq War.*[2]

The game follows a squad of U.S. Marines from 3rd Battalion, 1st Marines (3/1), fighting in the Second Battle of Fallujah over the span of six days in November 2004. The premise of the game was the subject of controversy in 2009, with questions raised as to its appropriateness, especially given the fact that the true events the game is based upon was recent at the time. It was originally to be published by Konami, however, in April 2009, a spokesman informed the Associated Press that Konami was no longer publishing the game due to the controversy surrounding it.*[3] As of September 2015, the game has not been released and there is no set release date.

2.42.1   Background

In an interview with Atomic Games president, Peter Tamte, he stated that "One of the divisions in our company was developing training tools for the United States Marine Corps, and they assigned some U.S. Marines from 3rd Battalion, 1st Marines to help us out." *[2] However, a few months into development, 3rd Battalion, 1st Marines (3/1) was deployed in Iraq and participated in the Second Battle of Fallujah.*[2] Tamte later stated that "When they came back from Fallujah, *they asked us* to create a videogame about their experiences there, and it seemed like the right thing to do." *[2] Tamte further stated that the goal of *Six Days in Fallujah* is to create the most realistic military shooter possible, and that "Ultimately, all of us are curious about what it would *really* be like to be in a war. I've been playing military shooters for ages, and at a certain point when I'm playing the game, I know it's fake. You can tell a bunch of guys sat in a room and designed it. That's always bothered me." .*[4] Tamte further elaborated in an interview with Joystiq that, "The words I would use to describe the game first of all, it's compelling. And another word I use insight. There are things that you can do in video games that you

cannot do in other forms of media. And a lot of that has to do with presenting players with the dilemmas that the Marines saw in Fallujah and then giving them the choice of how to handle that dilemma. And I think at that point, you know - when you watch a movie, you see the decisions that somebody else made. But when you make a decision yourself, then you get a much deeper level of understanding." .*[5] Tamte describes the project as "a meticulously recreated in-game version of Fallujah, complete with real life Marines lending their names and likenesses, as well as recreations of specific events from the battle. It's almost like time travel. You're experiencing the events as they really happened." *[6]

2.42.2   Development

The team at Atomic Games interviewed over 70 individuals, composed of the returning U.S. Marines, Iraqi civilians, Iraqi insurgents, war historians, and senior military officials, and learned the psychological complexity of the battle.*[6] The game's director, Juan Benito, elaborated that "Through our interviews with all of the Marines, we discovered that there was an emotional, psychological arc to the Battle of Fallujah.*[6]

Atomic Games describes *Six Days* as a survival horror game, but not in the traditional sense. The fear in *Six Days* does not come from the undead or supernatural, but from the unpredictable, terrifying, and real tactics employed by the insurgents that were scattered throughout Fallujah.*[7] Benito states that "Many of the insurgents had no intention of leaving the city alive, so their entire mission might be to lie in wait, with a gun trained at a doorway, for *days* just waiting for a Marine to pop his head in. They went door-to-door clearing houses, and most of the time the houses would be empty. But every now and then, they would encounter a stunningly lethal situation... which, of course, rattled the Marines psychologically." GamePro has stated that for Benito, giving players a taste of the horror, fear, and misery experienced by real-life Marines in the battle was a top priority. Benito states "These are scary places, with scary things happening inside of them. In the game, you're plunging into the unknown, navigating through darkened interiors, and 'surprises' left by the insurgency. In most modern military shooters, the tendency is to turn the volume up to 11 and keep it there. Our game turns it up to 12 at times but we dial it back down, too, so we can establish a cadence." *[7]

Atomic Games has also stated that the game's environments are 100% destructible and degradable thanks to a completely custom rendering engine, and it would surpass that of *Battlefield: Bad Company*.*[8] Tamte states that "This engine gives us more destructive capability than we've seen in any game, even games that aren't finished yet." According to the developers, destructible environments are critically important in telling the true story of the events in Fallujah, as the Marines eventually learned to blow holes in houses using C4, grenade launchers, and air strikes to blindside the insurgents waiting within, being considered as "combat puzzles".*[8] It is also stated that the claim of the game containing destructive environments is genuine and not based around a "goofy, out-of-place marketing gimmick." *[8]

On April 27, 2009, it was announced that, due to the controversial nature of the game, Konami suspended its role as a publisher. The game is still in development by Atomic games, but Konami will not be publishing it.*[9]

On August 6, 2009, Atomic Games said that they were unable to obtain a new publisher and would let go of some staff.*[10] A day later Industrygamers stated that they heard from a source, "Out of 75 people, less than a dozen are left and about a third of that isn't even developers. The remaining team is basically a skeleton cleanup crew that will be gone soon too. They are trying to downplay the extent of these layoffs, but the reality is that Atomic is pretty much dead." .*[11]

On March 2, 2010, IGN stated that the game is still coming out and is finished. However, as of January 2014, it has not been released.*[12]

In August 2012, it was revealed that Sony may have once considered publishing the title.*[13] Later that month, Atomic Games' president, Peter Tamte, informed the British website, Digital Spy, that *Six Days in Fallujah* was "definitely not canceled" and remains "very important" to the studio.*[14]

2.42.3 Controversy

Shortly after the announcement of the game, *Six Days* was met with criticism by British war veterans from the United Kingdom, as well as from a British peace group, Stop the War Coalition.*[15] Reg Keys, father of slain Royal Military Police Lance Corporal Thomas Keys, stated that "Considering the enormous loss of life in the Iraq War, glorifying it in a video game demonstrates very poor judgement and bad taste... These horrific events should be confined to the annals of history, not trivialized and rendered for thrill-seekers to play out... It's entirely possible that Muslim families will buy the game, and for them it may prove particularly harrowing. Even worse, it could end up in the hands of a fanatical young Muslim and incite him to consider some form of retaliation or retribution." *[16]

Tim Collins, a former lieutenant Colonel of the 1st Battalion Royal Irish Regiment, shared a similar disposition. Collins stated, "It's much too soon to start making video games

about a war that's still going on, and an extremely flippant response to one of the most important events in modern history. It's particularly insensitive given what happened in Fallujah, and I will certainly oppose the release of this game." *[16]

To counter the accusations made by critics, in an interview with Joystiq, Tamte stated that "As we've watched the dialog that's taken place about the game, there is definitely one point that we want people to understand about the game. And that is, it's not about the politics of whether the U.S. should have been there or not. It is really about the stories of the Marines who were in Fallujah and the question, the debate about the politics, that is something for the politicians to worry about. We're focused now on what actually happened on the ground." *[5]

2.42.4 References

[1] Gilbert, Ben (April 5, 2009). "Konami announces 'Six Days in Fallujah,' based on real battle in Iraq". Joystiq. Retrieved 2009-04-07.

[2] *GamePro* (#248). May 2009. p. 60. Missing or empty |title= (help)

[3] "Company pulls plug on `Fallujah' war video game". Associated Press.

[4] *GamePro*, Issue #248. May 2009 pg. 60-61

[5] Nelson, Randy (2009-04-13). "interview: Six Days in Fallujah". Joystiq. Retrieved 2012-07-24.

[6] *GamePro*, Issue #248. May 2009. pg. 61

[7] *GamePro*, Issue #248. May 2009. pg. 62

[8] *GamePro*, Issue #248. May 2009. pg. 63

[9] "News - Konami Drops Controversial Six Days in Fallujah". Gamasutra. Retrieved 2012-07-24.

[10] "Six Days in Fallujah Developer Cuts Staff - Xbox 360 News at IGN". Xbox360.ign.com. Retrieved 2012-07-24.

[11] "Rumor: Atomic Games at Death's Door?". IndustryGamers. 2009-08-07. Retrieved 2012-07-24.

[12] "Six Days in Fallujah Finished, Still Coming Out - Xbox 360 News at IGN". Au.xbox360.ign.com. Retrieved 2012-07-24.

[13] "Sony Considered Publishing Controversial 'Six Days in Fallujah' - PSLS". PSLS. Retrieved 2012-08-17.

[14] "Six Days in Fallujah 'definitely not canceled'". Digital Spy. August 26, 2012. Retrieved September 3, 2012.

[15] "Outrage Over Konami's "Six Days in Fallujah"". GamePolitics. 2012-07-20. Retrieved 2012-07-24.

[16] "Iraq War video game branded crass, insensitive". *Daily Mail* (London). April 7, 2009.

2.42.5 External links

- *Six Days in Fallujah* at IGN

2.43 Sonic X-treme

Not to be confused with Sonic Extreme.

Sonic X-treme was a cancelled platform video game in the *Sonic the Hedgehog* series. Developed by Sega Technical Institute (STI), *X-treme* was designed to capitalize on the success of Sega's mascot character by being the first fully 3D *Sonic* game and the first original *Sonic* title developed for the Sega Saturn. During the course of development, several different styles of gameplay were tried and the plot of the game changed several times.

Originally pitched as a two-dimensional platform game for the Sega Genesis, the game was eventually moved to development on the Saturn and for Windows, intended for release during the holiday season of 1996. However, *X-treme* became stuck in development hell after several incidents, including an unfavorable visit by Sega of Japan executives and issues with acquiring a game engine, made the deadline difficult to achieve. After two of the lead programmers for the project became ill, the game was eventually cancelled. Reviewers and video game journalists have retrospectively considered the possibility of what *Sonic X-treme* could have done for the Saturn had it been released, including comparisons to competing mascot video games *Super Mario 64* and *Crash Bandicoot*.

2.43.1 Premise

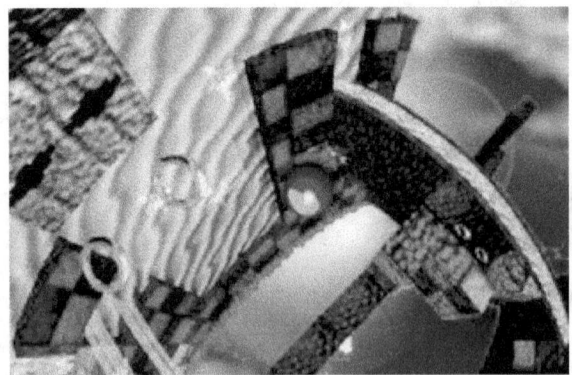

Jade Gully Zone, from Senn and Alon's engine

With the game constantly changing platforms, engines, and development teams, there were many loose storylines in consideration. According to developer Christian Senn,

about six or seven story lines were considered during the three-year development timeframe.[1] While originally based on the Saturday morning cartoon series,[2] the main storyline used in promotion of the final game in magazines involved a Professor Gazebo Boobowski and his daughter, Tiara. The two were the guardians of the six magical Rings of Order, as well as the ancient art of ring-smithing. Gazebo and Tiara feared that Dr. Robotnik was after the six Rings of Order, and called on Sonic to get the Rings before Robotnik could. Dr. Robotnik kidnapped Gazebo after he requested Sonic's help, making it so Sonic had to retrieve both him and the Rings of Order.[1] At one point in the development process, there was a possibility for 4 playable characters: Knuckles the Echidna, Tiara Boobowski, Miles "Tails" Prower, and Sonic the Hedgehog.[2] Other characters intended to be included in the game were Nack the Weasel and Metal Sonic, who would have been a boss character in the final level. Various moves were added to the characters, such as a ring toss move for Sonic, which was left out of development.[1]

To further the traditional "Sonic formula", every level was designed in a tube-like fashion; Sonic would be able to walk onto walls, thus changing the direction of gravity and the rotation of the level itself, much like the special stages in *Knuckles' Chaotix*. In addition, an unusual, fish-eye lens-styled camera was put into place so players could see more of their surroundings at any given time.[2]

> "3D Sonic is free to move around in a completely open 3D environment. Previously, on the 2D games, things were restricted to a very linear path, whereas now he can run around in the open without any restrictions to his path. The 360-degree rotation allows for new aspects to the gameplay. It means that Sonic can now do things like run from a wall onto the ceiling and explore lots of new hidden areas." [1]
> Executive Producer Mike Wallis

Senn has highlighted that the 3D gameplay still kept true to the *Sonic* series formula of collecting rings and speeding through game levels. Wallis also made mention to the game's overall layout, consisting of three acts per zone and varying in focuses on speed, exploration, and puzzle solving.[1]

2.43.2 Development

Following the completion of *Sonic & Knuckles* in 1994, Sega began working on the next game in its *Sonic the Hedgehog* franchise, which was known in development as *Sonic X-treme* and would have been the first *Sonic* game to feature fully 3D graphics. Development of the game was started by Sega Technical Institute, a U.S.-based developer that had worked on several previous *Sonic* games, beginning on the Sega Genesis and subsequently moving to the Sega 32X.[*][3] In its earliest conception, *Sonic X-treme* was designed for release on the Sega Genesis as a side-scrolling platform game, much like previous Sonic games for the system.[*][4] As new consoles and the beginning of the 32-bit era were on the way, the game was later moved to the Sega 32X and was known at this stage under the development names *Sonic 32X*[*][1] and *Sonic Mars*,[*][4] after the development name "Project Mars" used for the Sega 32X.[*][5] Even at this stage, the game's design changed wildly, including concepts such as an isometric viewpoint side-scroller. Eventually, however, development of the game was settled on a full 3D platform game.[*][4]

As the game's design had changed significantly[*][4] and evolved beyond the capabilities of the struggling Sega 32X, the game was shifted again to the Sega Saturn.[*][6] The Saturn version of the project was initially developed separately by two teams in parallel starting in the second half of 1995. One team, led by designer Chris Senn and programmer Ofer Alon, was in charge of developing the main game for PC.[*][6][*][7] The other team, led by Robert Morgan and including programmer Chris Coffin, worked on porting Senn and Alon's work to the Saturn while developing the "free-roaming, 'arena-style'" 3D boss engine.[*][6][*][6] Senn and Alon's "fixed-camera side-scroller" with the ability "to move freely in all directions" was similar to *Bug!*, and featured a fish-eye camera system (called the "Reflex Lens") that gave players a wide-angle view of the action.[*][4] As a result, levels appeared to move around Sonic.[*][6]

Screenshot from Coffin's "boss engine"

In March 1996, Sega of Japan representatives, including CEO Hayao Nakayama, visited STI headquarters to evaluate the game's progress. They were unimpressed by the main game engine's performance on the Saturn, although Senn and Alon did not have an opportunity to demonstrate the PC version.[*][6][*][7] Therefore, Nakayama requested the entire game be reworked around the boss engine.[*][7] To achieve this in time for the strict December 1996[*][6] deadline, Coffin's team was moved into a place of isolation from further company politics[*][6] and worked between sixteen[*][6] and twenty hours a day.[*][4] Then, in April, Bernie Stolar approached the STI team and inquired of Wallis what he could do to help the game meet its deadline. Wallis suggested that the game engine from Sonic Team's *Nights into Dreams...* would be helpful. Stolar agreed and acquired the engine. However, the engine's creator and lead programmer of the original Mega Drive Sonic games, Yuji Naka, reportedly threatened to leave the company if it was used.[*][4][*][7] STI lost two weeks of development time from the loss of the *Nights* engine.[*][1]

Senn and Alon had initially continued on with their game engine, undeterred by their work's original rejection, hoping to pitch it to Sega's PC division. However, it was eventually rejected again, prompting Alon to leave Sega.[*][4][*][6][*][7] No part of the STI team worked in unity during development.

> "We had artists doing art for levels that hadn't even been concepted out. We had programmers waiting and waiting and waiting until every minute detail had been concepted out, and we had designers doing whatever the hell they wanted. It was a mess, and because of the internal politics, it was even more difficult to get any work done." [*][1]
> Executive Producer Mike Wallis

By August, Chris Senn had become so ill that he was told he had six months to live though Senn would survive this ordeal and Chris Coffin came down with a severe case of pneumonia. With both Senn's team and Coffin's team crippled, Wallis was left with an incomplete *X-treme* and only two months before its deadline. At this point, he made the decision to cancel the game.[*][4] Although Sega initially stated that *X-treme* had merely been delayed,[*][8] the project was cancelled in early 1997.[*][7]

2.43.3 After cancellation

With the cancellation of *X-treme*, Sega instead decided to concentrate on a port of the Genesis title *Sonic 3D Blast*, and Sonic Team's *Nights into Dreams...* for the 1996 holiday season.[*][6] Sonic Team started work on an original 3D *Sonic* title for the Saturn, which eventually became *Sonic Adventure* for the Dreamcast. According to Naka, remnants of the project can be seen in the compilation game *Sonic*

With X-treme'*s cancellation, the Sega Saturn had no original Sonic the Hedgehog platform game released for the console*

Jam.[9][10] STI was officially disbanded in 1996 as a result of changes in management at Sega of America.[3]

For many years, very little content from the game was ever released beyond screenshots that had been released to the media in promotion of the game prior to its cancellation. However, in 2006 a copy of a very early test engine was sold at auction to an anonymous collector who bought it for US$2500.[11] An animated GIF image of the gameplay was initially released, and the disk image itself was leaked on July 17, 2007 after a fundraising project by the "Assemblergames" website community purchased the disc from the collector.[12] In 2006, Chris Senn opened the "*Sonic X-treme* Compendium" web site and began revealing large amounts of the game's development history to the public, including videos of early footage, a playable character named Tiara, and a large amount of previously unreleased concept music related to the title. He also was given permission by Hirokazu Yasuhara, the level designer for the majority of the original 16-bit *Sonic* titles, to post level designs that were going to be put in the game. Senn, along with the community, announced intentions to recreate the game,[2] but ultimately the project was canceled in January 2010.[13]

In early 2015, fans from *Sonic* fansite "Sonic Retro" obtained the game's source code, polished it up into a playable build, and released it for download on the internet.[14] It is largely faithful to the game's original state, with their changes merely being made to piece together code to make it playable and running on modern computers.[14] The first release only features one level, the Jungle-themed level from the 1996 E3 promotional video, but the team expects to be able to release more levels in the future.[14]

2.43.4   Legacy

The *Sonic X-treme* debacle has been cited as a reason for the ultimate failure of the Sega Saturn.[5][15] With the *Sonic the Hedgehog* series being attributed to much of the suc-

cess of the company's prior system, the Genesis, and Sony and Nintendo both having flagship 3D platformers available early in the life cycle of their consoles (*Crash Bandicoot* and *Super Mario 64*, respectively), Sega was expected by fans to follow suit and produce an official 3D *Sonic* game.[1] With the game's cancellation, the Saturn never did receive an exclusive *Sonic* platform game, but rather only the Genesis port of *Sonic 3D Blast; Sonic Jam*, a compilation of the 2D Genesis *Sonic* titles; and *Sonic R*, a racing game. Sonic's debut in a full 3D platform game was not until 1998, with *Sonic Adventure* as a Dreamcast launch title, well after the discontinuation of the Saturn.[2]

Following the game's cancellation, journalists and fans have speculated about the impact a completed *X-treme* might have had on the market. David Houghton of GamesRadar described the prospect of "a good 3D *Sonic* game" on the Saturn as "a 'What if...' situation on a par with the dinosaurs not becoming extinct." [6] IGN's Travis Fahs called *X-treme* "the turning point not only for SEGA's mascot and their 32-bit console, but for the entire company", although he also noted that the game served as "an empty vessel for SEGA's ambitions and the hopes of their fans" .[4] Dave Zdyrko, who operated a prominent website for Saturn fans during the system's lifespan, offered a more nuanced perspective: "I don't know if [*X-treme*] could've saved the Saturn, but ... *Sonic* helped make the Genesis and it made absolutely no sense why there wasn't a great new *Sonic* title ready at or near the launch of the [Saturn]".[16] In a 2013 retrospective, producer Mike Wallis maintained that *X-treme* "definitely would have been competitive" with Nintendo's *Super Mario 64.*[7] Websites such as Destructoid and GamesRadar have speculated the game could have been a source of inspiration for future games such as 2007's *Super Mario Galaxy.*[2][6] Several journalists would also note similarities between *X-treme* and the 2013 game *Sonic Lost World.*[17][18]

2.43.5   References

[1] Allen, Jonathan. "Spotlight: Sonic X-treme" . Lost Levels. Retrieved 2012-07-23.

[2] Davis, Ashley (2008-11-19). "What could have been: Sonic X-treme" . Destructoid. Retrieved 2012-07-23.

[3] Horowitz, Ken (June 11, 2007). "Developer's Den: Sega Technical Institute" . Sega-16. Retrieved 2014-04-16. **Roger Hector:** When it became obvious that Sony was taking the lead, Sega's corporate personality changed. It became very political, with lots of finger-pointing around the company. Sega tried to get a handle on the situation, but they made a lot of mistakes, and ultimately STI was swallowed up in the corporate turmoil.

[4] Fahs, Travis (2008-05-29). "Sonic X-Treme Revisited - Saturn Feature at IGN" . IGN. Retrieved 2012-07-23.

[5] Newton, James (2011-06-23). "Feature: The Sonic Games That Never Were". Nintendo Life. Retrieved 2012-07-23.

[6] Houghton, David (April 24, 2008). "The greatest Sonic game we never got ...". GamesRadar. Retrieved 2012-07-23.

[7] "The Making Of: Sonic X-treme". Edge Online. 2013-04-14. Retrieved 2014-06-15.

[8] "New Sega Happenings". Next Generation Online. Archived from the original on 1996-12-20. Retrieved 2014-05-04. *GunBlade NY* and *Sonic X-treme* have now both been officially scheduled for Saturn release in 1997 ... [*X-treme*] had previously been scrapped to be reworked.

[9] Barnholt, Ray. "Yuji Naka Interview: Ivy the Kiwi and a Little Sega Time Traveling". 1UP.com. Archived from the original on 2014-09-07. Retrieved 2014-03-04.

[10] Towell, Justin (2012-06-23). "Super-rare 1990 Sonic The Hedgehog prototype is missing". GamesRadar. Retrieved 2014-03-04. **Yuji Naka:** The reason why there wasn't a Sonic game on Saturn was really because we were concentrating on NiGHTS. We were also working on *Sonic Adventure* that was originally intended to be out on Saturn, but because Sega as a company was bringing out a new piece of hardware the Dreamcast we resorted to switching it over to the Dreamcast, which was the newest hardware at the time. So that's why there wasn't a Sonic game on Saturn. With regards to *X-treme*, I'm not really sure on the exact details of why it was cut short, but from looking at how it was going, it wasn't looking very good from my perspective. So I felt relief when I heard it was cancelled.

[11] Snow, Blake (2006-03-09). "Man pays $2500 for Sonic X-treme demo". Joystiq. Retrieved 2012-07-23.

[12] McWhertor, Michael (2007-06-04). "Sonic X-Treme "Nights Version"". Kotaku.com. Retrieved 2012-07-23.

[13] Senn, Christian (2010-01-12). "Game Forums @ Senntient.com / Public Announcement Regarding Project-S". Senntient.com. Retrieved 2012-07-23.

[14] http://www.eurogamer.net/articles/2015-02-24-sonic-fans-release-long-lost-tech-demo-of-unfinished-saturn-game

[15] Buchanan, Levi (2009-02-02). "What Hath Sonic Wrought? Vol. 10 - Saturn Feature at IGN". IGN. Retrieved 2012-07-23.

[16] Sewart, Greg (2005-08-05). "Sega Saturn: The Pleasure And The Pain". 1UP.com. Archived from the original on 2012-10-21. Retrieved 2014-03-17.

[17] Ponce, Tony (2013-05-28). "Sonic Lost World trailer reminds me of Sonic X-treme". Destructoid. Retrieved 2013-08-01.

[18] Sliwinski, Alexander (2013-05-28). "Sonic: Lost World finds gameplay footage". Joystiq. Retrieved 2013-08-01.

2.44 Stacked Volumetric Optical Disk

The **Stacked Volumetric Optical Disc** (or SVOD) is an optical disc format developed by Hitachi/Maxell, which uses an array of wafer-thin optical discs to allow data storage.

Each "layer" (a thin polycarbonate disc) holds around 9.4 GB[*][1] of information, and the wafers are stacked in layers of 100 or so, giving overall data storage increase of 100× or more.

SVOD might likely be a candidate, along with HVDs, to be a next-generation optical disc standard.

2.44.1 References

[1] "New DVD Discs Will Store One Terabyte Data". RealityPod. Retrieved 21 November 2012.

2.44.2 External links

- Hitachi Maxell develops wafer-thin storage disc details and interview from IDG News Service (4 October 2006)

- Maxell details in Japanese language (19 April 2006)

- NikkeiBP some details in English (20 April 2006)

2.45 StarCraft: Ghost

StarCraft: Ghost is a cancelled military science fiction stealth-action video game previously under development by Blizzard Entertainment. Part of Blizzard's *StarCraft* series, the game was announced on September 20, 2002, and was to be developed by Nihilistic Software for the Nintendo GameCube, Xbox, and PlayStation 2 video game consoles. Several delays in development caused Blizzard to move back the release date and the game has not yet materialized. Nihilistic Software ceded development to Swingin' Ape Studios in 2004 before Blizzard bought the company, and plans for the GameCube version were cancelled in 2005.

Blizzard announced in March 2006 that the game was put on "indefinite hold" while the company investigated seventh generation video game console possibilities. Subsequent public statements from company personnel had been contradictory about whether production was to be renewed or planned story elements worked into other products. The continued delay of *Ghost* had caused it to be labeled as vaporware, and it was ranked fifth in *Wired News* ' annual

Vaporware Awards in 2005. In 2014, Blizzard president Mike Morhaime confirmed that *Ghost* was officially cancelled.

Unlike its real-time strategy predecessor *StarCraft*, *Ghost* is a third-person shooter, and was intended to give players a closer and more personal view of the *StarCraft* universe. Following Nova, a Terran psychic espionage operative called a "ghost", the game is set four years after the conclusion of *StarCraft: Brood War* and covers a conspiracy about a secretive military project conducted by Nova's superiors in the imperial Terran Dominion. Very little of the game's storyline has been released; however, in November 2006 after the game's postponement, a novel was published called *StarCraft Ghost: Nova*, which covers the backstory of the central character.

2.45.1   Gameplay

Campaign

A screenshot of the game in 2005 just prior to its postponement. Nova is shown engaging a group of Terran guards in a firefight.

During *StarCraft: Ghost* 's gameplay, the player's character Nova must use stealth and darkness to reach objectives and remain undetected. Nova has a cloaking device that allows for temporary concealment, but certain hostile nonplayer characters can overcome this with special devices and abilities.[1] Nova is also equipped with thermal imaging goggles and a special EMP device for disabling electronic devices and vehicles. In addition to the focus on stealth elements, *StarCraft: Ghost* includes a complex combat system. Blizzard planned to include a small arsenal of weaponry with assault and sniper rifles, grenades, shotguns, and flamethrowers.[2] Nova can engage in hand-to-hand combat and uses these skills to eliminate enemy threats quietly. If alerted, enemy characters will hunt for the player,

set up traps, and fire blindly to nullify Nova's cloaking device.[1]

Nova is highly agile, acrobatic, and able to perform maneuvers such as mantling and climbing ledges, hanging from pipes, and sliding down ziplines.[1] The player has access to Nova's psionic powers honed through training as a ghost agent, such as the ability to improve her speed and reflexes drastically.[3] *StarCraft: Ghost* includes many of the vehicle units featured in *StarCraft* and *StarCraft: Brood War*. Some vehicles, such as space battlecruisers and starfighters, only play support roles, while others, such as hoverbikes, scout cars, and futuristic siege tanks, can be piloted by the player.[4]

Multiplayer

The multiplayer mode in *StarCraft: Ghost* differs from the stealth-based mechanics of the single-player portion. It aims to give players a personal view of the battles from the real-time strategy games of the series. Accordingly, *Ghost* 's multiplayer is structured around class-based team gameplay and fighting in a variety of game modes. *Ghost* incorporates traditional game modes from multiplayer video games such as deathmatch, capture the flag, and king of the hill, but also introduces two game modes specifically designed for the *StarCraft* universe. The first is "Mobile Conflict", which requires two teams to fight for control of a single Terran military factory with the ability of atmospheric flight. Using vehicles and team tactics, both teams must first board the structure and then capture its control room to fly it to the team's starting point. The structure must then land and be defended from capture by the opposing team for a set amount of time.[5]

The second unique game mode is "Invasion", in which two teams fight for control of mineral resource nodes. Whenever teams capture a node they gain points that can be used to purchase classes and vehicles.[6] In all of the team-based game modes, teams have access to four Terran unit classes: light infantry, marine, firebat, and ghost. The light infantry class has minimal armor but a larger range of weapons,[7] while the marine is a heavily armored soldier with an assault rifle and grenades.[8] The firebat is a heavy marine armed with a flamethrower and napalm rockets.[9] Finally, the ghost is a variation of Nova's character in the single-player mode, equipped with a cloaking device, thermal vision, EMP device, and sniper rifle, but lacks the speed ability.[10] Due to the size of the armor worn by marines and firebats, only ghosts and light infantry can pilot vehicles.[11]

2.45.2 Plot

Ghost takes place in the fictional universe of the *StarCraft* series. The series is set in a distant part of the galaxy called the Koprulu Sector and begins in the year 2499. Terran exiles from Earth are governed by a totalitarian empire, the Terran Dominion, that is opposed by several smaller rebel groups. Two alien races discover humanity: the insectoid Zerg, who begin to invade planets controlled by the Terrans; and the Protoss, an enigmatic race with strong psionic power that attempt to eradicate the Zerg.*[12] *Ghost* takes place four years after the conclusion of *StarCraft: Brood War*, in which the Zerg become the dominant power in the sector and leave both the Protoss and the Dominion in ruins.*[13] The game follows the story of Nova, a young ghost agent a human espionage operative with psychic abilities in the employ of the Dominion.

Nova, the game's protagonist, appears in a cinematic from Ghost. *The cinematics were designed to be of higher quality than those in previous* StarCraft *titles.*

Although the game has been indefinitely postponed, the backstory for Nova was released in the novel *StarCraft Ghost: Nova* by Keith R. A. DeCandido. It was meant to accompany the game's release, but was published in 2006 after development halted.*[14] In the novel, Nova is a fifteen-year-old girl and daughter to one of the ruling families of the Confederacy of Man, an oppressive government featured in *StarCraft*. The Confederacy is overthrown by rebels, who go on to form the Dominion. Nova has significant psionic potential, but has been kept out of the Confederate ghost operative training program because of her father's influence. After her family is murdered by rebels, Nova loses control of her mental abilities and accidentally kills 300 people around her home. She flees from her home before she is caught, and is later forced to work for an organized crime boss as an enforcer and executioner. She is rescued by a Confederate agent who is investigating her disappearance during a rebel attack on the Confederate capital that leads to the Confederacy's destruction. Nova is consequently acquired by the newly formed Terran Dominion, who erase her memory and train her as a ghost agent.*[15]

Few details have been revealed about *Ghost* 's plot beyond Nova's backstory. Under emperor Arcturus Mengsk, the Terran Dominion has rebuilt much of its former strength and controls a new military formed to counter the Zerg. To further bolster the effectiveness of his military, Mengsk initiates a secret research operation codenamed Project: Shadow Blade and places it under the command of his right-hand man, General Horace Warfield. In the program, an experimental and potentially lethal gas called terrazine is used to enhance the genetic structure of the Dominion's psychic ghost agents. The process is described as changing the agents into "shadowy superhuman beings bent on executing the will of their true master". It is into the midst of this that Nova finishes her training and is dispatched in operations against the Koprulu Liberation Front, a rebel group that challenges Mengsk's empire. However, Nova's mission leads her to uncover a conspiracy that involves Shadow Blade. This revelation causes her to question her loyalty to the Dominion and could upset the balance of power within the galaxy.*[16]

2.45.3 Development

On September 20, 2002, Blizzard Entertainment announced the development of *StarCraft: Ghost* in conjunction with fellow video game company Nihilistic Software.*[17] Nihilistic aimed to release the game for the Xbox, PlayStation 2 and Nintendo GameCube video game consoles in late 2003, which elicited positive reactions from the press.*[18] The game was consistently delayed, and during the third quarter of 2004, Nihilistic discontinued their work on the project.*[19] Blizzard stated that Nihilistic had completed the tasks it had been contracted for, and the game would be delivered on time.*[20]

In July 2004, Blizzard Entertainment began collaboration with Swingin' Ape Studios to work on the game,*[21] and bought the company in May 2005.*[22] Despite anticipation for the game by industry journalists,*[23]*[24] *Ghost* was delayed again and its release date was pushed back to September 2005. At Electronic Entertainment Expo 2005, *Ghost* was officially reannounced,*[25] but the GameCube version was canceled by Swingin' Ape Studios due to the platform's lack of online support.*[26] The game's release was again delayed until 2006. Despite the efforts of Swingin' Ape, *Ghost* failed to materialize as scheduled, and in March 2006 Blizzard Entertainment announced an indefinite postponement on development of *Ghost* while the company explored new options with the emerging seventh generation of video game consoles.*[27] Despite its long development history, IGN noted that the concept of *Ghost* still held promise.*[28] Although the game's development was suspended, Keith R. A. DeCandido's novel *StarCraft Ghost: Nova* was published several months later in November 2006.*[29]

Complementing Nihilistic's and Swingin Ape Studio's work on the game, Blizzard's cinematics team originally formed to develop *StarCraft* 's cut scenes*[30] created the cut scenes for *Ghost* 's single-player campaign, which are integral to the game's storyline. The team, which originally consisted of six people, grew to 25, and used newer hardware, software, and cinematics techniques to create higher quality cut scenes than those featured in *StarCraft* and *Brood War*.*[31] The game's trailer, composed of the cinematics team's work, was released in August 2005.*[32]

Since *Ghost* 's production halted, Blizzard Entertainment has sporadically released information about the title. The game's protagonist, Nova, shows up in one campaign mission of *StarCraft II: Wings of Liberty*, in which players are given the option to side with her or fight against her forces. She also features in *StarCraft II: Heart of the Swarm*. Metzen further stated that he believed *Ghost* had an excellent storyline that may be told in future novels following from DeCandido's *Nova*.*[33] In June 2007, Rob Pardo, one of the lead developers at Blizzard Entertainment, indicated that there still was interest in finishing *Ghost*.*[34] Later in an interview, Pardo stated that Blizzard had been "stubborn" in persevering with *Ghost*, but they "were not able to execute [the game] at the level we wanted to" .*[35] Blizzard's president Mike Morhaime and Pardo gave a presentation on the company's history at the D.I.C.E. Summit in February 2008. During the presentation, they listed games canceled by Blizzard, which did not include *Ghost*. When questioned about this, Blizzard's co-founder Frank Pearce explained that the title was never "technically canceled" and that it was not in the company's focus at the time due to a finite amount of development resources.*[36] Morhaime later elaborated that it was the sudden success of *World of Warcraft* and the concurrent development of *StarCraft II* that consumed Blizzard's resources, leading to *Ghost* being put on hold.*[37] Despite Blizzard's announcements, many of the video games industry's journalists now list *Ghost* as canceled and consider it vaporware; the game was ranked fifth in the 2005 edition of *Wired News* ' annual Vaporware Awards.*[38]*[39]*[40]

On September 23, 2014 in an interview with *Polygon* about the cancellation of Blizzard's next generation MMO *Titan*, Mike Morhaime confirmed that *StarCraft: Ghost* was also cancelled. Morhaime said, "It was hard when we canceled *Warcraft Adventures*. It was hard when we canceled *StarCraft: Ghost*, but it has always resulted in better-quality work." *[41]

2.45.4   References

[1] Remo, Chris (2005-11-05). "*StarCraft: Ghost* Preview: single-player" . Shacknews. Archived from the original on 2005-11-27. Retrieved 2008-04-11.

[2] "Covert Ops: Weapons" . *StarCraft: Ghost*. Blizzard Entertainment. Archived from the original on 2006-02-06. Retrieved 2008-04-11.

[3] "Covert Ops: Psi Powers" . *StarCraft: Ghost*. Blizzard Entertainment. Archived from the original on 2006-02-06. Retrieved 2008-04-11.

[4] "Covert Ops: Vehicles" . *StarCraft: Ghost*. Blizzard Entertainment. Archived from the original on 2006-02-06. Retrieved 2008-04-11.

[5] "Multiplayer" . *StarCraft: Ghost*. Blizzard Entertainment. Archived from the original on 2006-02-06. Retrieved 2008-04-11.

[6] Remo, Chris (2005-11-11). "*StarCraft: Ghost* Preview: Multiplayer" . Shacknews. Retrieved 2008-04-11.

[7] "Multiplayer Characters: Light Infantry" . *StarCraft: Ghost*. Blizzard Entertainment. Archived from the original on 2006-02-06. Retrieved 2008-04-11.

[8] "Multiplayer Characters: Marine" . *StarCraft: Ghost*. Blizzard Entertainment. Archived from the original on 2006-02-06. Retrieved 2008-04-11.

[9] "Multiplayer Characters: Firebat" . *StarCraft: Ghost*. Blizzard Entertainment. Archived from the original on 2006-02-06. Retrieved 2008-04-11.

[10] "Multiplayer Characters: Ghost" . *StarCraft: Ghost*. Blizzard Entertainment. Archived from the original on 2006-02-06. Retrieved 2008-04-11.

[11] 1UP Staff (2005-10-28). "StarCraft Ghost (Xbox)". 1UP.com. Retrieved 2008-10-05.

[12] "The Story So Far: Part 1: *StarCraft*". Blizzard Entertainment. 2007-11-21. Retrieved 2007-11-22.

[13] "The Story So Far: Part 2: The Brood War" . Blizzard Entertainment. 2008-04-16. Retrieved 2008-04-16.

[14] Metzen, Chris (May 2006). "Introduction" . *StarCraft Ghost: Nova*. Simon & Schuster. pp. v–vii.

[15] "Covert Ops: Nova Backstory" . *StarCraft: Ghost*. Blizzard Entertainment. Archived from the original on 2006-02-06. Retrieved 2008-04-15.

[16] "Covert Ops: Story" . *StarCraft: Ghost*. Blizzard Entertainment. Archived from the original on 2006-02-06. Retrieved 2008-04-15.

[17] "Press Release" . *StarCraft: Ghost*. Blizzard Entertainment. 2002-09-20. Archived from the original on 2002-10-04. Retrieved 2008-04-16.

[18] "TGS 2002: StarCraft: Ghost Impressions". IGN. 2002-09-20. Retrieved 2008-04-23.

[19] Adams, David (2004-06-22). "Nihilistic Exits *StarCraft: Ghost*". IGN. Retrieved 2008-04-16.

[20] "FAQ". *StarCraft: Ghost*. Blizzard Entertainment. 2004. Archived from the original on 2004-07-01. Retrieved 2008-04-16.

[21] Van Autrijve, Rainier (2004-07-07). "Blizzard Taps Swingin' Ape to work on *StarCraft: Ghost*". GameSpy. Retrieved 2008-04-16.

[22] Vasconcellos, Eduardo (2004-05-16). "Blizzard Gets a New Monkey on its Back". GameSpy. Retrieved 2008-04-16.

[23] "E3 2004: StarCraft Ghost". IGN. 2004-05-11. Retrieved 2008-04-23.

[24] Polak, Steve (2004-02-26). "*Ghost* rider in the sky". *The Courier-Mail* (Queensland, Australia: News Corporation). p. 8.

[25] Clayman, David (2005-05-18). "E3 2005: *StarCraft Ghost* Returns". IGN. Retrieved 2008-04-16.

[26] Thorsen, Tor (2005-11-03). "*StarCraft: Ghost* not beaming onto GameCube". GameSpot. Retrieved 2008-04-16.

[27] "Blizzard Postpones *StarCraft: Ghost* Indefinitely". GameSpy. 2006-03-24. Retrieved 2008-04-16.

[28] "*StarCraft: Ghost* Goes To Heaven?". IGN. 2006-03-24. Retrieved 2008-04-24.

[29] "*StarCraft Ghost: Nova* (Mass Market Paperback)". Simon & Schuster. Archived from the original on 2012-06-19. Retrieved 2007-12-02.

[30] "Joeyray: Blizzard Movie-Making". *10th Anniversary Celebration*. Blizzard Entertainment. Archived from the original on 2001-04-18. Retrieved 2008-01-08.

[31] "Interview with the *StarCraft: Ghost* Cinematics Team". *StarCraft: Ghost*. Blizzard Entertainment. 2003-03-12. Archived from the original on 2003-03-12. Retrieved 2008-10-06.

[32] "StarCraft: Ghost Videos". IGN. 2005-08-25. Retrieved 2008-10-06.

[33] "Starcraft Panel Discussion: Lore". GameSpot. 2007-08-08. Retrieved 2008-10-05.

[34] "Blizzard Still Has Hope For *StarCraft: Ghost*". Slashdot. 2007-06-28. Retrieved 2007-11-29.

[35] "Blizzard Still Has Hope For *StarCraft: Ghost*". *Edge*. 2007-06-27. Retrieved 2008-10-09.

[36] Totilo, Stephen (2008-02-13). "Blizzard Explains Why *StarCraft: Ghost* Wasn't On The DICE Canceled Games List". MTV Multiplayer. Retrieved 2008-04-15.

[37] Crecente, Brian (2011-02-14). "The Head of Blizzard Explains the Death of StarCraft Ghost". Kokatu. Retrieved 2011-09-17.

[38] "*StarCraft: Ghost*". IGN. Retrieved 2008-04-15.

[39] "*StarCraft: Ghost*". GameSpot. Retrieved 2008-04-15.

[40] Kahney, Leander (2006-02-06). "Vaporware: Better Late Than Never". *Wired News*. Retrieved 2008-04-15.

[41] Philip, Kollar (2014-09-23). "Blizzard cancels its next-gen MMO Titan after seven years". *Polygon*. Retrieved 2015-07-02.

2.45.5 See also

- *Warcraft Adventures: Lord of the Clans*, another cancelled Blizzard project

2.46 Stonekeep

Stonekeep is a role-playing video game developed and released by Interplay Entertainment for the PC in 1995. It is a first-person dungeon crawler game with pre-rendered environments, digitized characters and live-action cinematic sequences. Repeatedly delayed, the game that was supposed to be finished in nine months eventually took five years to make, the record longest development of a video game at the time. The game featured the voice of Arthur Burghardt - well known as the character Destro in the *G.I. Joe* 1985 cartoon series - in the role of an evil god named Khull-Khuum.

2.46.1 Gameplay

Stonekeep *screenshot showing the Ice Queen boss fight at the Ice Caverns level*

Stonekeep is a first-person RPG in the style of *Eye of the Beholder* and *Dungeon Master*. The game is set in a series of underground labyrinths, filled with monsters, treasures and traps. The player uses the directional keys on the keyboard for movement and typing in notes in the journal and uses the mouse pointer to interact with objects and characters. The mouse pointer is usually a target indicator for aiming attacks and weapons wherever it is clicked. When the mouse pointer is moved onto a particular something it changes to another icon to indicate a different action. For example, the mouse pointer changes to an eye when the player can examine things (often signs) or the mouse pointer changes into a spread-out hand when the player can pick up items. Other mouse pointers include opening and closing chests, opening and closing panels, pulling levers and switches, pressing buttons, drinking water and giving items

The protagonist Drake has two starting possessions: the magic scroll and the magic mirror. The magic scroll allows the player to pick up an infinite number of items. Items of the same type can be combined together up to a maximum quantity of 99; other items can be combined together such as a quiver which can hold 99 arrows. The magic mirror allows the player to equip Drake and other characters with weapons, armour and accessories and to consume items to affect their status such as healing potions or bad smelling Throg food otherwise Drake can read scrolls used on him. Although Drake can wield any weapon, other characters like Farli and Karzak can only wield hammers, axes and shields. Certain weapons like polearms and heavy swords require Drake to have two free hands to wield one. Some armour can be worn by certain characters. For instance only dwarves can wear dwarven platemail and only Drake can wear knight armour. Exceptional characters like Sparkle and Wahooka cannot be outfitted, but can still consume items.

The third possession is the journal available once the player procures it. The journal is divided into six sections. The first section of the journal records the statistics of Drake, shows the status of his current equipped weapons and describes the characteristics of his partners. Drake statistics are strength, agility, health and his weaponry skills including polearm, sword, magick, missiles and others. The second section of the journal records any clues and hints the player may come across. The third section of the journal is used for writing notes. The fourth section of the journal records items each time the player picks up a new type. The fifth section of the journal records runes each time the player comes across a new one. Unlike the items section, the runes do not have their own respective name recorded. The sixth section of the journal records the level maps that the players journeys through. Spots on the map can be clicked and notes referring to them can be written.

Stonekeep features an elaborate 'Magick' system where four types of runes are inscribed onto a spellcaster (wand): Mannish, Fae, Throggish, and Meta. The first three runes are used for offensive, defensive and special interaction pur-

poses. The Meta runes enhance the effectiveness of the base runes, like double power multiplies a single firebolt in two. To use a spellcaster, it must contain adequate mana and runes must inscribed onto the shaft of the spellcaster. To inscribe runes on the spellcaster, the player needs to equip the spellcaster on either one of Drake's hands, open the journal onto the Runes section, take out the spellcaster and copy the runes onto the spell slots of the spellcaster. Then the runecaster can be taken out at any time and the spell that can be used has the spell slot highlighted and launched on the indicated target point. Using the magic mirror, some spells can be aimed on the characters especially healing and quickness spells.

2.46.2 Plot

Stonekeep's mythology revolves around a variety of gods associated with planets of the solar system. In order, they are Helion (Mercury), Aquila (Venus), Thera (Earth), Azrael (Mars), Marif (Jupiter), Afri (Saturn), Saffrini (Uranus), Yoth-Soggoth (Neptune) the Master of Magick, and Kor-Soggoth (Pluto) the Brother to Magick. These gods were captured and imprisoned in nine orbs by the dark god Khull-Khuum 1000 years before the events of the game, during a cataclysm referred to as "The Devastation".

Stonekeep is centered on a hero, Drake. Ten years before the events of the game, Drake's home, the castle of Stonekeep, was destroyed by the insane god Khull-Khuum, the Shadowking. Drake, at this time just a boy, was saved from the castle by a mysterious figure. Returning to the ruins of Stonekeep, Drake is visited by the goddess Thera, who sends his spirit out of his body into the ruins itself to explore, find the mystical orbs containing the other gods, and reclaim the land.

Along the way, Drake makes many friends, including Farli, Karzak, and Dombur the dwarves; the great dragon Vermatrix; the elf Enigma; and the mysterious Wahooka, the King of goblins. Together, they embark on a quest of ridding the world of Khull-Khuum and his evil minions and allies.

2.46.3 Development

The project started out with just two people, Peter Oliphant and Michael Quarles. It was intended to last only nine months and only supposed to cost $50K. However, because the initial stages of the game looked good it exceeded nine months, lasting a total of five years. Eventually there was a production crew of over 200 people, and costing a total of $5 million. The intro sequence was the most expensive part of the production, costing nearly half a million dollars to produce, which was ten times more than the initial budget for the entire project.

The initial story line was written by Oliphant, who also designed and programmed the graphics and artificial intelligence engine for the game. The project started out being called *Brian's Dungeon* (named after Brian Fargo, the president of Interplay Entertainment at the time). Fargo came up with the final name, *Stonekeep*. The production took much longer than expected because of the rapid advancement of personal computer hardware at the time; specifically, PC CPUs advancing from 80386, to 80486, to Pentiums in the years the game was being developed. Oliphant, who originally designed the game and was lead programmer, left the game as the project passed its fourth year in development. He felt his continued presence was resulting in the constant addition of feature creep and changes (he was a contractor, and had initially only signed up for a nine-month project). After he left, the design became finalized and the product was shipped one year later. Quarles, who was an Interplay employee, stayed as the game's producer and saw it through to the end.

The initial specification for the game included that it could *not* require a hard drive or a mouse, run on an 80286 CPU, use 640K, and run off floppy disks. At the project's end, the game had been upgraded to *requiring* a mouse, a hard drive, a 386 CPU, and ran off CD-ROM. As a result, the engine had to be extensively modified throughout the production. The initial motions of the monsters in the game were captured by using a blue screen outside with the sunlight. This resulted in uneven lighting from take to take, so eventually all that work was scrapped. Later a professional studio with controlled lighting was used.

According to Oliphant,*[2] when the project was taken over by Quarles, two questionable decisions were made. The game was always designed to be grid-based, where the player moved from grid to grid (in contrast to today's full freedom of motion 3D environments). Oliphant wanted the movement from center of grid to center of grid, but Quarles changed this to edge of grid to edge of grid. This resulted in the problem that turning within a grid moved the player to the other side of the grid. Much of the long production was a result of correcting this lack of symmetry. The other questionable decision was to not include Oliphant in the production of the motion graphics (Oliphant had extensive Hollywood background before becoming a game developer). One consequence was that the original combat graphics had been captured from the waist up only, as Quarles had reasoned one must be close to a monster to fight it. Peter Oliphant, upon being delivered these graphics and seeing them for the first time, pointed out that the player could back away during a fight, which would result in seeing their legs. The legs therefore had to be drawn in by hand frame-by-frame to fix this, until these graphics were scrapped for a professional green screen treatment used later on. The original skeleton in the game was an actual skeleton being worn by one of the artists, and was filmed against a green screen. Because of this there were no images/animations of the skeleton walking away from the player during game play. A few months before the game's release the skeleton was replaced with the 3D model which was used on the packaging.

About three years into the project, Oliphant suggested to Fargo that the product be delivered on CD-ROM. Fargo rejected this idea at the time, citing the failure of previous Interplay CD-ROM projects that had gone this route. Oliphant suggested this after Fargo requested him to drop his percentage of royalties by half due to the high cost of production and goods to create the product, as it was at that time to be shipped on eight floppy disks. The cost of one CD was about the cost of one floppy disk, and the possibilities for eight floppy disks having problems is much greater than a single CD, so the solution seemed obvious to Oliphant. And, in fact, six months later Fargo changed his mind and made the same decision.

2.46.4 Release

Stonekeep was originally released for the PC DOS and Windows 95 in 1995, packaged in an elaborate gravestone-style illustrated box and came with a white hardback novella *Thera Awakening*, coauthored by Steve Jackson and David L. Pulver (all rights of the novel went to Interplay). The CD-ROM also included a file called "muffins.txt" which contained a recipe for "Tim Cain's Chocolate Chip Pumpkin Muffins". Years later, *Stonekeep* was later made available for purchase through GOG.com's digital distribution system for Windows XP and Windows Vista.*[1]

2.46.5 Reception

Maximum praised the game's "atmospheric 3D rendered world" and sound effects, but criticized the lack of challenging puzzles, low amount of gore, and "sluggish" combat. They gave it 3 out of 5 stars.*[3] In June 1996, *Computer Gaming World* named *Stonekeep* as the role-playing game reader's choice winner in its Game of the Year awards.*[4] In 2000, The Adrenaline Vault stated it is "the best game of 1995 in its genre and is still an awesome game to play. You're guaranteed to spend hours upon hours on this game and never tire of its thrill!"*[5] A 1996 review by Bernard Yee of GameSpot did not offer similar praises, concluding that "*Stonekeep* is a dated first-person RPG that suffers from a poor interface, little depth, and few frills." *[6]

In 1996, editors of *Computer Gaming World* ranked it as the tenth top vaporware title in computer game history (due 1991, delivered 1996), stating that "after seeing the same basic demo for years, the game finally shipped, as an anti-climax." *[7] *Stonekeep* was also ranked at number six on GameSpot's top ten vaporware hall of shame.*[8] In 2009, GamesRadar also included *Stonekeep* among the games "with untapped franchise potential" due to the cancelation of *Stonekeep 2: Godmaker*.*[9]

2.46.6 Legacy

The Oath of Stonekeep, a novel set in the world of *Stonekeep*, was written by Troy Denning and published 1999 by Berkley Boulevard Books.*[10]

Interplay's Black Isle Studios worked on a sequel, *Stonekeep 2: Godmaker*,*[11] for roughly five years, before eventually cancelling it in 2001 in order to work on *Icewind Dale II* and *Baldur's Gate III: The Black Hound*.

The game, alongside a novelization, would remain the only entry in its series until the 2010 announcement of *Stonekeep: Bones of the Ancients*, a game developed for Interplay by Alpine Studios. It is not a sequel to *Stonekeep*, but rather an all new game and a standalone entry in the franchise.*[12] *Bones of the Ancients* was released in 2012 as downloadable content at WiiWare.

2.46.7 References

[1] "*Stonekeep* purchase page". GOG.com. Retrieved 2009-09-12.

[2] "Matt Chat 136: Peter Oliphant on Stonekeep". Youtube.com. 2012-02-11. Retrieved 2013-08-24.

[3] "Maximum Reviews: Stonekeep". *Maximum: The Video Game Magazine* (Emap International Limited) (2): 159. November 1995.

[4] Wilson, Johnny; et al. (June 1996). "The Computer Gaming World 1996 Premier Awards: Role-Playing Game of the Year". *Computer Gaming World* (143): 58.

[5] Clair, Brian (2000-01-01). "*Stonekeep* Review". The Adrenaline Vault. Retrieved 2009-09-12.

[6] Yee, Bernard (1996-05-08). "*Stonekeep* Review". GameSpot. Retrieved 2009-09-12.

[7] Wilson, Johnny; et al. (November 1996). "15 Top Vaporware Titles in Computer Game History". *Computer Gaming World* (148): 130.

[8] Poole, Stephen. "Vaporware Hall of Shame". GameSpot. Archived from the original on 2009-05-16. Retrieved 2009-09-12.

[9] 123 games with untapped franchise potential, GamesRadar US, April 30, 2009

[10] "Amazon Reader entry of the novel "The Oath of Stonekeep"". Amazon.com. 1999-10-01. Retrieved 2013-08-24.

[11] "Stonekeep 2: Godmaker". GameSpot.com. Retrieved 2013-08-24.

[12] JC Fletcher (2010-01-13). "Interplay resurrecting Stonekeep on WiiWare". Joystiq. Retrieved 2013-08-24.

2.46.8 External links

- *Stonekeep* at MobyGames

2.47 Team Fortress 2

"TF2" redirects here. For other uses, see TF2 (disambiguation).

Team Fortress 2 is a team-based first-person shooter multiplayer video game developed by Valve Corporation. It is the sequel to the 1996 mod *Team Fortress* for *Quake* and its 1999 remake. It was released as part of the video game compilation *The Orange Box* on October 10, 2007 for Windows and the Xbox 360.*[3] A PlayStation 3 version followed on December 11, 2007.*[2] On April 8, 2008, it was released as a standalone title for Windows. The game was updated to support OS X on June 10, 2010, and Linux on February 14, 2013. It is distributed online through Valve's download retailer Steam; retail distribution was handled by Electronic Arts.

In *Team Fortress 2*, players join one of two teams comprising nine character classes, battling in a variety of game modes including capture the flag and king of the hill. The development is led by John Cook and Robin Walker, creators of the original *Team Fortress*. Announced in 1998, the game once had more realistic, militaristic visuals and gameplay, but this changed over the protracted nine-year development. After Valve released no information for six years, *Team Fortress 2* regularly featured in *Wired News* ' annual vaporware list among other ignominies.*[4] The finished *Team Fortress 2* has cartoon-like visuals influenced by the art of J. C. Leyendecker, Dean Cornwell and Norman Rockwell*[5] and is powered by Valve's Source engine.

Team Fortress 2 received critical acclaim for its art direction, gameplay, humor, and use of character in a multiplayer-only game.*[6]*[7]*[8][9] Valve continues to release new content, including maps, items and game modes. On June 23, 2011, it became free to play, supported by microtransactions for unique in-game equipment. A 'drop system' was also added and refined in this update, allowing free-to-play users to still receive game equipment by use of a random number generator.

2.47.1 Gameplay

Team Fortress 2 *in play; a group of RED players attack a BLU base on the map* "Well"

Like its predecessors, *Team Fortress 2* is focused around two opposing teams competing for a combat-based principal objective.*[10] In the game's fiction, the teams are composed of mercenaries hired by two feuding brothers to protect the company assets belonging to one brother while trying to destroy those of the other; the teams are thus represented by the names of these companies: Reliable Excavation & Demolition (RED) and Builders League United (BLU). Players can choose to play as one of nine character classes in these teams, each with his own unique strengths, weaknesses, and weapons. Although the abilities of a number of classes have changed from earlier *Team Fortress* incarnations, the basic elements of each class have remained, that being one primary weapon, one secondary weapon, and one melee weapon.*[11]*[12] The game was released with six official maps, although 44 extra maps, 9 arena maps, 8 king of the hill maps, and various other map types have been included in subsequent updates.*[13]*[14] In addition, a number of community assembled maps have been released. When players join a level for the first time, an introductory video shows how to complete its objectives. During matches, "The Administrator" ,*[15] a woman voiced by Ellen McLain, announces various game events over loudspeakers.*[16] The player limit is 16 on the Xbox 360 and PlayStation 3.*[17] On the PC, in 2008 Valve updated *Team Fortress 2* to include a server variable that allows up to 32 players.*[18]

Team Fortress 2 is the first of Valve's multiplayer games to provide detailed statistics for individual players. They include: time spent playing as each class, most points obtained, and the most captures or objectives achieved in a single life. Persistent statistics tell the player how he or she is improving in relation to these statistics, such as if a player

comes close to his or her record for the damage inflicted in a round.*[13] *Team Fortress 2* also features numerous "achievements" for carrying out certain tasks, such as scoring a certain number of kills or completing a round within a certain time. New sets of class-specific achievements have been added in updates, which add new abilities and weapons to each class once unlocked by the player. This unlockable system has since been expanded into a random-chance system, where the player can also obtain the items simply by playing the game.*[19]

Game modes

The objective of the game is defined by the game mode in use.

- In Capture the Flag maps, the objective for both teams is to obtain a briefcase of intelligence from the enemy team's base and return it to their own base while preventing the opposing team from doing the same. The player carrying the intelligence can be killed to drop the briefcase, or the player can willingly drop it; in either case this starts a two-minute timer. If the intelligence is not collected by another player on the opposing team before that timer expires, it is returned to its home base. A team can only score by delivering the enemy's intelligence to their base. A match lasts until one team scores a set number of points or time runs out.*[20]

- A variant of Capture the Flag, called *Special Delivery*, has teams vying over a single suitcase in an attempt to deliver to the end target. Once the intelligence is picked up by one team, only members of that team can carry it until either it successfully delivered, or the drop timer expires and it is returned to its original spawn point. Like with Territorial Control, there exist's only one official map (called Doomsday) for this game mode, though unlike Territorial Control, a *halloween-themed* variant exists for Doomsday, named *Doomsday Event.*

- Control Point modes are more varied in their objectives, but share the common aim of capturing a particular point on the map by having one or more team members stay on the point without the presence of the opposing team for a short period of time.*[20] In standard control point maps, each team already controls an equal number of points, with one additional point left unclaimed; teams can only attempt to capture points that are nearest to those points they already control. Each team attempts to progressively capture all the control points to win the round. In attack/defend-style maps, the RED team already controls all the points on the map, and must hold these points from being captured by the opposing team (BLU Team) for a length of time. Attack/defend-style maps can come in either single-round or multiple-round maps; Rounds past the first will only be played if BLU wins a round..

- Territorial control is a more complex version of Control Point, in which each map is divided into a number of closed sections, held between the two teams. Each round has one team attempting to seize control of the opposing team's capture point for that section in a fixed amount of time. Once a team holds all but the enemy's home base point, they then attempt to capture the enemy base point while the enemy defends for a fixed amount of time.*[20] The only official map made by Valve for Territorial Control is called *Hydro*, presumably because of its complex design..

- King of the Hill features a single control point in the center of the map, with teams attempting to gain control over it. The control point is locked for a set amount of time after the round start. Once the point is unlocked, either team can capture it. Capturing the control point starts the team's clock counting down. The clock that is currently counting down will only stop if the other team captures the point, the clock reaches 0:00, or another map-specific event occurs such as a Halloween boss spawning during the Halloween event. Teams win the round when their clock reaches 0:00 with the control point currently under their control. If the other team is capturing the point, the round will not end until the point's capture progress resets to 0.*[21]

- Payload maps feature a length of track atop which sits a bomb mounted on a cart. This mode have two variants; Payload and Payload Race. In normal Payload maps, one team must escort the cart through a series of checkpoints in a limited amount of time to reach a final target that the other team defends. The cart moves when one or more players of the team are near it without enemies nearby; during this, the cart heals and provides ammo to those close by. If no one is by the cart, it may move backwards towards the last checkpoint, or roll off uphill sections of the track.*[22] In Payload Race, both teams race to deliver a bomb-cart to a final target; there are no checkpoints and unmanned carts will not move in reverse, but still require team members nearby to fully complete uphill sections.*[23] Both types of Payload maps can come in single-round or multiple-round maps. In standard payload, further rounds will only be played if BLU wins. In Payload Race, all rounds will be played regardless

of which team wins, but the winner of previous rounds starts with their cart farther up the track.

- Arena is a team deathmatch mode, typically occurring smaller environments featuring a single control point. In each round, players do not respawn on death; a team wins the round by capturing the control point or eliminating all of the other team members.*[24]

- Mann vs. Machine (shortened to MvM) is a six player co-operative mode where the mercenaries are defending a structure against increasingly difficult waves of robots. Players have the ability to buy upgrades and improvements between rounds using in-game money earned during previous waves.*[25] A "Mann Up" version is available by purchasing tickets with real money to participate in larger events on official servers with the opportunity to win unique cosmetic items after successfully completing a Tour of Duty.*[26]

A "Training mode" exists to help new players get used to the game, using wooden target dummies for practice and to introduce them to concepts of the multiplayer modes. Game modes allowing respawn all have a delay before a player may respawn again, with the respawn system being set up in waves of 10 seconds. Since 2009, there is also typically a Halloween-themed variation on one or more of the above modes during the weeks around the holiday, with maps or modes updated to include themed decorations and often a more difficult challenge to the players. For example, for Halloween 2012, this included an extremely difficult Mann vs Machine round involving destroying more than 800 enemy forces.*[27] Due to popular demand of the Halloween events, Valve added Full Moon, the ability to play these events on the few days around the full moon throughout the year, and later Eternaween, the ability for players to vote to play the themed-events for a two-hour block at any other time.*[28]

Classes and characters

Team Fortress 2 features nine playable classes, categorized into offense, defense, and support roles.*[13] Each class has strengths and weaknesses, and must work with other classes to be efficient, encouraging strategy and teamwork.*[29] Each class has at least three weapons: a unique primary weapon, a common or unique secondary weapon, and a unique melee weapon.

Offense

- The **Scout** (voiced by Nathan Vetterlein) is a cocky, fast-talking baseball fan and street runner from

From left to right: Pyro, Engineer, Spy, Heavy, Sniper, Scout, Soldier, Demoman, Medic

Boston, Massachusetts, who practiced running to "beat his maddog siblings to the fray." *[30] He is a fast, agile character, armed with a scattergun, a pistol and an aluminum baseball bat. The Scout can double-jump and captures control points and pushes payloads twice as fast as other classes, but has the lowest health.

- The **Soldier** (Rick May) is more durable, but slower. A jingoistic American military man (despite the fact that he was never actually in the Army),*[31] the Soldier is armed with a rocket launcher, shotgun, and a folding shovel. The Soldier can use his rocket launcher to rocket jump to higher positions at the cost of some health.

- The **Pyro** (Dennis Bateman) is a mentally unstable pyromaniac of unknown gender or origin, clad in a fire-retardant suit and a voice-muffling gas mask. As well as a shotgun and fire axe, the Pyro is armed with a homemade flamethrower which can set players on fire and produce a blast of compressed air that knocks away nearby enemies and projectiles and extinguishes burning teammates.*[32]*[33]

Defense

- The **Demoman** (Gary Schwartz) is a black, one-eyed, heavy-drinking Scotsman from Ullapool.*[34] Armed with a glass bottle, grenade and sticky bomb launchers, the Demoman can use his explosives to provide indirect fire and set traps.*[32]

- The **Heavy** (Gary Schwartz) is a large Russian from the USSR, heavy in stature and accent, and obsessed with firepower. Though he is the slowest class, he can sustain and deal immense amounts of damage. His default weapons consist of his fists, a shotgun, and an enormous minigun that he affectionately refers to as "Sasha" .*[35]

- The **Engineer** (Grant Goodeve) is a relaxed and intellectual "good ol' boy" from Bee Cave, Texas.*[36] The Engineer can build structures to support his team: a sentry gun for defending key points, a health and ammunition dispenser, and a teleporter.*[32] He is armed with a standard shotgun and pistol, and a wrench which is also used to repair, upgrade, and speed up building of his structures. He can also remotely destroy his structures.

Support

- The **Medic** (Robin Atkin Downes) is a German doctor from Stuttgart with little regard for the Hippocratic Oath.*[37] The Medic's "Medi Gun" heals teammates and gradually builds "ÜberCharge"; on activation, the ÜberCharge grants boosts such as temporary invulnerability to the Medic and patient. The Medic is also equipped with an air-powered syringe gun and bonesaw.*[22]*[32] He keeps doves as pets, one of which is named Archimedes.

- The **Sniper** (John Patrick Lowrie) is a cheerful New Zealand ocker-style character*[38] equipped with a laser-sighted sniper rifle to shoot enemies from afar and a submachine gun and a kukri for close combat.*[32]

- The **Spy** (Bateman) is a French double agent equipped with covert tools, including a cloaking device disguised as a watch, an electronic sapper used to sabotage and potentially destroy enemy Engineers' buildings, and a device hidden in his cigarette case that enables him to disguise as enemy players. Armed with a revolver, the Spy can also use his butterfly knife to stab enemies in the back or sides (known as a backstab), which instantly kills them.*[32]

Non-playable characters   Other characters include the Administrator (voiced by Ellen McLain), an unseen announcer who provides information about time limits and objectives to players, and her assistant, Miss Pauling (Ashly Burch). The cast has expanded with Halloween updates, including the characters of Merasmus, the Bombinonicon, and the Mann Brothers (all voiced by Nolan North).

Competitive play

Team Fortress 2 is played competitively, through multiple leagues. The North American league ESEA supports a paid *Team Fortress 2* league, with $10,080 in prizes for the top teams in 2012.*[39] In addition, *Team Fortress 2* is played in multiple free-to-play leagues including UGC (United Gaming Clans), ETF2L in Europe.*[40]*[41] and AsiaFortress in Asia*[42] *Team Fortress 2* is played competitively mostly in one of three gamemodes: Highlander (one of each class, 9 players per team), 6v6 (2 Scouts, 2 Soldiers, 1 Demoman, and 1 Medic with other classes used in certain situations), or 4v4 (1 Scout, 1 Soldier, 1 Demoman, and 1 Medic, with other classes used often). While formalized competitive gameplay is very different from normal *Team Fortress 2*, it offers an environment with a much higher level of teamwork than in public servers (also known as "pubs"). Most teams use a VOIP program to communicate, and use a combination of strategy, communication, and aiming ability to win against other teams. Many competitive leagues also feature an item banlist, in order to speed up gameplay and remove untested or undesirable strategies from matches. Valve has revealed they plan to implement a matchmaking system based on these two modes, similar to that of *Counter-Strike: Global Offensive.**[43]

2.47.2 Development

Origins

Team Fortress originally began life as a free mod for *Quake*. Development on *Team Fortress 2* switched to the GoldSrc engine in 1998 after the development team Team Fortress Software consisting of Robin Walker and John Cook were first contracted and finally outright employed by Valve Corporation.*[44] At the point of Team Fortress Software's acquisition production moved up a notch and the game was promoted to a standalone, retail product;*[44] to tide fans over, work began on a simple port of the game which was released in 1999 as the free *Team Fortress Classic.**[45] Notably, *Team Fortress Classic* was built entirely within the publicly available *Half-Life* Software Development Kit as an example to the community and industry of its flexibility.*[46]

Walker and Cook had been heavily influenced by their three-month contractual stint at Valve, and now they were working full-time on their design, which was undergoing rapid metamorphosis. *Team Fortress 2* was to be a modern war game, with a command hierarchy including a commander with a bird's-eye view of the battlefield, parachute drops over enemy territory, networked voice communication and numerous other innovations.*[47]

Early development

The new design was revealed to the public at E3 1999, where it earned several awards including Best Online Game and Best Action Game.*[48] By this time *Team Fortress 2*

The game's visual style changed drastically over its development.

had gained a new subtitle, *Brotherhood of Arms*, and the results of Walker and Cook working at Valve were becoming clear. Several new and at the time unprecedented technologies on show: Parametric animation seamlessly blended animations for smoother, more lifelike movement,*[49] and Intel's multi-resolution mesh*[49] technology dynamically reduced the detail of on-screen elements as they became more distant to improve performance*[49] (a technique made obsolete by decreasing memory costs; today games use a technique known as level of detail, which uses more memory but less processing power). No release date was given at the exposition.

In mid–2000, Valve announced that development of *Team Fortress 2* had been delayed for a second time.*[50] They attributed the delay to development switching to an in-house, proprietary engine that is today known as the Source engine. It was at around this time that all news ran dry and *Team Fortress 2* entered six years of silent development.*[51] During that time, both Walker and Cook worked on various other Valve projects Walker was project lead on *Half-Life 2: Episode One*[52] and Cook became a Steam developer.*[53]

Final design

The next significant public development occurred in the run up to *Half-Life 2* 's 2004 release: Valve's director of marketing Doug Lombardi claimed that *Team Fortress 2* was still in development and that information concerning it would come after *Half-Life 2* 's release. This did not happen; nor was any news released after Lombardi's similar claim during an early interview regarding *Half-Life 2: Episode One.**[54] Before *Episode Two* 's release Gabe Newell again claimed that news on *Team Fortress 2* would be forthcoming, and *Team Fortress 2* was re-unveiled a month later at the July 2006 EA Summer Showcase event.*[11]

Both teams sport their own art style to help players navigate the levels.

Walker revealed in March 2007 that Valve had quietly built "probably three to four different games" before settling on their final design.[*][55] Due to the game's lengthy development cycle it was often mentioned alongside *Duke Nukem Forever*, another long-anticipated game that had seen many years of protracted development and engine changes.[*][4] The beta release of the game featured six multiplayer maps, of which three contain optional commentary by the developers on the game design, level design and character design, and provide more information on the history behind the development.[*][56]

Team Fortress 2 does not attempt the realistic graphical approach used in other Valve games on the Source engine such as *Half-Life 2*, *Counter-Strike: Source* and *Day of Defeat: Source*. Rather, it uses a more stylized, cartoon-like approach "heavily influenced by early 20th century commercial illustrations" and achieved with extensive use and manipulation of phong shading.[*][12][*][57] The development commentary in the game suggests that part of the reason for the cartoonish style was the difficulty in explaining the maps and characters in realistic terms. The removal of an emphasis on realistic settings allows these explanations to be sidestepped.[*][56] The game debuted with the Source engine's new dynamic lighting, shadowing and soft particle technologies, among many other unannounced features, alongside *Half-Life 2: Episode Two*. *Team Fortress 2* was also the first game to implement the Source engine's new Facial Animation 3 features.[*][58]

The art style for the game was inspired by J. C. Leyendecker, as well as Dean Cornwell and Norman Rock-

well.[*][5] Their distinctive styles of strong silhouettes and shading to draw attention to specific details were adapted in order to make the models distinct, with a focus on making the characters' team, class and current weapon easily identifiable. Silhouettes and animation are used to make the class of a character apparent even at range, and a color scheme that draws attention to the chest area brings focus to the selected weapon.[*][59] The voices selected for each of the classes were based on imagining what people from the 1960s would expect the classes to have sounded like, according to writer Chet Faliszek.[*][60]

The map design has a strong evil genius theme with archetypical spy fortresses, concealed within inconspicuous buildings such as industrial warehouses and farms to give plausibility to their close proximities; these bases are usually separated by a neutrally themed space. The bases hide exaggerated super weapons such as laser cannons, nuclear warheads, and missile launch facilities, taking the role of objectives. The maps have little visual clutter and stylized, almost impressionistic modeling, to allow enemies to be spotted more easily. The impressionistic design approach also affects textures, which are based on photos that are filtered and improved by hand, giving them a tactile quality and giving *Team Fortress 2* its distinct look. The bases are designed to let players immediately know where they are. RED bases use warm colors, natural materials and angular shapes, while BLU bases use cool colors, industrial materials and orthogonal shapes.[*][59]

Release

During the July 2006 Electronic Arts press conference, Valve revealed that *Team Fortress 2* would ship as the multiplayer component of *The Orange Box*. A conference trailer showcasing all nine of the classes demonstrated for the first time the game's whimsical new visual style. Managing director of Valve Gabe Newell said that the company's goal was to create "the best looking and best-playing class-based multiplayer game".[*][11] A beta release of the entire game was made on Steam on September 17, 2007 for customers who had pre-purchased *The Orange Box*, who had activated their *Black Box* coupon, which was included with the ATI HD 2900XT Graphics cards, and for members of the Valve Cyber Café Program.[*][61][*][62] The beta continued until the game's final release.

The game was released on October 10, 2007, both as a standalone product via Steam and at retail stores as part of *The Orange Box* compilation pack, priced at each gaming platform's recommended retail price. *The Orange Box* also contains *Half-Life 2*, *Half-Life 2: Episode One*, *Half-Life 2: Episode Two*, and *Portal*. Valve offered *The Orange Box* at a ten percent discount for those who pre-purchased it via

Steam before the October 10, as well as the opportunity to participate in the beta test.*[63]

Post-release

Since the release of *Team Fortress 2*, Valve has continually released free updates and patches through Steam for Microsoft Windows, OS X, and Linux users; though most patches are used for improving the reliability of the software or to tweak gameplay changes, several patches have been used to introduce new features and gameplay modes, and are often associated with marketing materials such as comics or videos offered on the *Team Fortress 2* website; this blog is also used to keep players up to date with the ongoing developments in *Team Fortress 2*.*[64] As of July 2012, each class has been given a dedicated patch that provides new weapons, items, and other gameplay changes; these class patches typically included the release of the class's "Meet the Team" video.*[33]*[65]*[66] Other major patches have included new gameplay modes including the Payload,*[22] Payload Race,*[23] Training,*[67] Highlander,*[68] Medieval,*[69] and Mann vs Machine modes.*[25] Themed patches have also been released, such as a Halloween-themed map with unique items available only during a set period around the holiday.*[70] Other new features have given players the ability of crafting new items within the game,*[71] trade items with other players,*[72] purchase in-game items through funds in Steam,*[73] and save and edit replay videos that can be posted to YouTube.*[74]

Valve has released tools to allow users to create maps, weapons, and cosmetic items through a contribution site; the most popular are added as official content for the game.*[75]*[76] This approach has subsequently created the basis for the Steam Workshop functionality of the software client.*[77]*[78] In one case, more than fifty users from the content-creation community worked with Valve to release an official content update in May 2013, with all of the content generated by these players.*[79] Valve reported that as of June 2013, over $10 million has been paid back to over 400 community members that have helped to contribute content to the game, including a total of $250,000 for the participants in the May 2013 patch.*[80] To help promote community-made features, Valve has released limited time events, such as the "Gun Mettle" or "Invasion" events in the second half of 2015, in which players can spend a small amount of money which is paid back to the community developers for the ability to gain unique items offered while playing on community-made maps during the event.*[81]*[82]*[83] *[84]

Development of the new content has been confirmed for the Xbox 360, while development for the PlayStation 3 was deemed "uncertain" by Valve.*[85] However, the PlayStation 3 version of *Team Fortress 2* received an update that repaired some of the issues found within the game, ranging from graphical issues to online connectivity problems; this update was included in a patch that also repaired issues found in the other games within *The Orange Box*.*[86] The updates released on PC and planned for later release on Xbox 360 include new official maps and game modes, as well as tweaks to classes and new weapons that can be unlocked through the game's achievement system.*[87] The developers attempted to negotiate with Xbox 360 developer Microsoft to keep the Xbox 360 releases of these updates free,*[88] but Microsoft refused and Valve announced that they would release bundles of several updates together to justify the price.*[89]

On June 10, 2010, *Team Fortress 2* was released for OS X, shortly after the release of Steam for OS X.*[67] The release was teased by way of an image similar to early iPod advertising, showing a dark silhouette of the Heavy on a bright green background, his Sandvich highlighted in his hand.*[90] Virtual earbuds, which can be worn when playing on either OS X or Windows once acquired, were given to players playing the game on OS X before June 14,*[91] though the giveaway period was later extended to August 16.*[92]

On November 6, 2012, Valve announced the release of *Team Fortress 2* for Linux as part of a restricted beta launch of Steam on the platform.*[93] This initial release of Steam and *Team Fortress 2* was targeted at Ubuntu with support for other distributions planned for the future. Later, on December 20, 2012, Valve opened up access to the beta, including *Team Fortress 2*, to all Steam users without the need to wait for an invitation.*[94] On February 14, 2013, Valve announced the full release of *Team Fortress 2* for Linux.*[95] From then to March 1, anyone who played the game on Linux would receive a free Tux penguin, which can be equipped in-game.

Team Fortress 2 was announced in March 2013 to be the first game to officially support the Oculus Rift, a consumer-grade virtual reality headset. A patch will be made to the client to include a "VR Mode" that can be used with the headset on any public server.*[96]

Trading   In *Team Fortress 2* players can trade for items such as weapons, cosmetics and utilities. Weapons and utilities can change and affect gameplay in different ways, the main being; some weapons have different stats than others and thus allows a different playstyle. Cosmetic items, on the other hand, do not change game-play at all as all they do is change the way the player looks.

In late 2011, the gaming site Kotaku reported that the *Team Fortress 2* 's trading economy was worth over $50 million.[*][97]

Free-to-play   On June 23, 2011, Valve announced that *Team Fortress 2* would become free to play. Unique equipment including weapons and outfits would be available as microtransactions through the in-game store, tied through Steam.[*][98] Walker stated that Valve would continue to provide new features and items free.[*][98] Walker stated that Valve had learned that the more players *Team Fortress 2* had, the more value it had for each player.[*][98]

The move came a week after Valve introduced several third-party free-to-play games to Steam and stated were working on a new free-to-play game.[*][99] Within nine months of becoming free to play, Valve reported that revenue from *Team Fortress 2* had increased by a factor of twelve.[*][100]

2.47.3   Marketing

The Scout talks about himself in his entry into the "Meet the Team" series.

To promote the game, Valve has released an ongoing video advertisement series entitled "Meet the Team" since May 2007. Constructed using the game engine and slightly more detailed character models, the series consists of short videos on individual characters, displaying their personalities and tactics. The videos are usually interspersed with clips of the character in combat in the game. The manners which these are presented have varied drastically: the first installment, "Meet the Heavy", depicted an interview with the gun-obsessed Eastern European[*][35] while "Meet the Soldier" showed the Soldier giving a misinformed lecture on Sun Tzu to a collection of severed heads as if to raw recruits.[*][31] The videos are generally released through Valve's services, though in one notable exception, the "Meet the Spy" video was leaked on YouTube during the Sniper/Spy update week.[*][101][*][102] The "Meet the Team" videos are based on the audition scripts used for the voice actors for each of the classes; the "Meet the Heavy" scripts is nearly word-for-word a copy of the Heavy's script. More recent

videos, such as "Meet the Sniper", contain more original material.[*][103] The videos have been used by Valve to help improve the technology for the game, specifically improving the facial animations, as well as a source of new gameplay elements, such as the Heavy's "Sandvich" or the Sniper's "Jarate" .[*][103] The final video in the Meet the Team series, "Meet the Pyro", was released on June 27, 2012.[*][104][*][105] Newell has stated that Valve is using the "Meet the Team" shorts as a means of exploring the possibilities of making feature film movies themselves. Newell believed that only game developers themselves have the ability to bring the interesting parts of a game to a film, and suggested that this would be the only manner through which a *Half-Life*-based movie would be made.[*][106] A fifteen-minute short, titled "Expiration Date", was released on June 17, 2014.[*][107] The shorts were made using Source Filmmaker, which was officially released and has been in open beta as of July 11, 2012.[*][108]

In more recent major updates to the game, Valve has presented teaser images and online comic books that expand the fictional history of the *Team Fortress 2*, as part of the expansion of the "cross-media property", according to Newell.[*][109] In August 2009, Valve brought aboard American comic writer Michael Avon Oeming to teach Valve "about what it means to have a character and do character development in a comic format, how you do storytelling" .[*][109] "Loose Canon", a comic associated with the Engineer Update, establishes the history of RED versus BLU as a result of the last will and testament of Zepheniah Mann in 1890, forcing his two bickering sons Blutarch and Redmond to vie for control of Zepheniah's lands between them; both have engineered ways of maintaining their mortality to the present, waiting to outlast the other while employing separate forces to try to wrest control of the land.[*][110] This and other comics also establish other background characters such as Saxton Hale, the CEO of Mann Co., the company that provides the weapons for the two sides and was bequeathed to one of Hale's ancestors by Zepheniah, and the Administrator, the game's announcer, that watches over, encourages the RED/BLU conflict, and keeps each side from winning.[*][111] The collected comics were published by Dark Horse Comics in *Valve Presents: The Sacrifice and Other Steam-Powered Stories*, a volume along with other comics created by Valve for *Portal 2* and *Left 4 Dead*, and released in November 2011.[*][112] Cumulative details in updates both in-game and on Valve's sites from 2010 through 2012 were part of a larger alternate reality game preceding the reveal of the Mann vs Machine mode, which was revealed as a co-op mode on August 15, 2012.[*][25][*][113][*][114]

Valve had provided other promotions to draw players into the game. Valve has held weekends of free play for *Team Fortress 2*.[*][115] Through an early update, hats and accessories can be changed or added to any of the classes, giving players some ability to customize the look of their character. New weapons were added in updates to allow the player to choose a loadout that best suit the player. Hats and weapons can be gained as a rare random drop, through the crafting / trading systems, or via cross-promotion: Limited-edition hats and weapons have been awarded for pre-ordering or gaining Achievements in other content from Steam, both from Valve (such as *Left 4 Dead 2*[*][116][*][117] and *Alien Swarm*) or other third-party games such as *Sam & Max: The Devil's Playhouse*, *Worms Reloaded*, *Killing Floor*, or *Poker Night at the Inventory* (which features the Heavy class as a character). According to Robin Walker, Valve introduced these additional hats as an indirect means for players to show status within the game or their affiliation with another game series simply by visual appearance.[*][118] The Red Pyro, Heavy, and Spy all function as a single playable character in the PC release of *Sonic and All-Stars Racing Transformed*.[*][119] The game's first television ad premiered during the first episode of the fifth season of *The Venture Bros.* in June 2013, featuring in-game accessories that were created with the help of Adult Swim.[*][120]

2.47.4 Reception

See also: Critical reception of The Orange Box

Upon release, *Team Fortress 2* received widespread critical acclaim, with overall scores of 92/100 and 92.60%, respectively on Metacritic and GameRankings.[*][122][*][121] Many reviewers praised the cartoon-styled graphics, and the resulting light-hearted gameplay,[*][8] and the use of distinct personalities and appearances for the classes impressed a number of critics, with *PC Gamer UK* stating that "until now multiplayer games just haven't had it." [*][9] Similarly, the game modes were received well, *GamePro* described the settings as focusing "on just simple fun",[*][128] while several reviewers praised Valve for the map "Hydro" and its attempts to create a game mode with variety in each map.[*][7][*][9] Additional praise was bestowed on the game's level design, game balance and teamwork promotion.[*][6] *Team Fortress 2* has received several awards individually for its multiplayer gameplay[*][129][*][130] and its graphical style,[*][131][*][132][*][133] as well as having received a number of "game of the year" awards as part of *The Orange Box*.[*][134][*][135]

Although *Team Fortress 2* was well received, *Team Fortress 2* 's removal of class-specific grenades, a feature of previous *Team Fortress* incarnations, was controversial amongst reviewers. IGN expressed some disappointment over this,[*][7] while conversely *PC Gamer UK* approved, stating "grenades have been removed entirely thank God"

.[*][9] Some further criticism came over a variety of issues, such as the lack of extra content such as bots[*][7] (although Valve have since added bots in an update[*][136]), problems of players finding their way around maps due to the lack of a minimap, and some criticism over the Medic class being too passive and repetitive in his nature.[*][9] The Medic class has since been re-tooled by Valve, giving it new unlockable weapons and abilities.

2.47.5 References

[1]   "Team Fortress 2" . *Steam*. Valve Corporation. Retrieved August 10, 2010.

[2]   "The Orange Box" . Metacritic. Retrieved November 20, 2014.

[3]   "Orange Box Goes Gold" . Joystiq. September 27, 2007. Retrieved November 20, 2014.

[4]   "Vaporware: Better Late Than Never" . Wired News. February 6, 2006. Archived from the original on January 10, 2009. Retrieved May 23, 2007.

[5]   Mitchell, Jason; Francke, Moby; Eng, Dhabih (August 6, 2007). "Illustrative Rendering in *Team Fortress 2*" (PDF). Valve Corporation. Retrieved August 10, 2007. Video summary (WMV, 75.4MB)

[6]   Gerstmann, Jeff (October 11, 2007). "*The Orange Box* Review" . GameSpot. Retrieved July 8, 2007.

[7]   Onyett, Charles (October 9, 2007). "*Team Fortress 2* Review" . IGN. Retrieved May 2, 2008.

[8]   Wong, Steven (October 12, 2007). "*Team Fortress 2* Review" . GameDaily. Archived from the original on October 13, 2007. Retrieved May 2, 2008.

[9]   Francis, Tom (October 10, 2007). "PC Review: *Team Fortress 2*". *PC Gamer UK*. ComputerAndVideoGames.com. Archived from the original on October 11, 2007. Retrieved May 2, 2008.

[10] "*Meet the Team*". *Steam*. Valve Corporation. Retrieved January 29, 2009.

[11] "*Half-Life 2: Episode Two* – The Return of *Team Fortress 2* and Other Surprises" . GameSpot. July 13, 2006. Retrieved July 8, 2015.

[12] Berghammer, Billy (March 28, 2007). "*Team Fortress 2* Hands-On Preview" . Game Informer. Archived from the original on April 6, 2007. Retrieved April 13, 2007.

[13] Berghammer, Billy (March 27, 2007). "The *Team Fortress 2* Interview: The Evolution" . Game Informer. Archived from the original on April 6, 2007. Retrieved April 13, 2007.

[14] "*Team Fortress 2* Badlands preview" . Shacknews. January 14, 2008. Retrieved January 21, 2008.

[15] "Team Fortress 2 – The Administrator". Valve Corporation. December 9, 2009. Retrieved December 29, 2011.

[16] "Ellen McLain". IMDB. Retrieved July 26, 2009.

[17] "*Team Fortress 2* Interview". IGN. April 10, 2007. Retrieved August 19, 2007.

[18] "*Team Fortress 2* February 28, 2008 Team Fortress 2 update". Valve Corporation. February 28, 2008. Retrieved October 16, 2009.

[19] Francis, Tom (January 22, 2008). "Team Fortress 2 Gets Unlockable Weapons". *PC Gamer UK*. Computer and Video Games. Archived from the original on January 24, 2008. Retrieved March 2, 2008.

[20] Bramwell, Tom (May 22, 2007). "*Team Fortress 2* First Impressions". Eurogamer. Retrieved May 23, 2007.

[21] "Team Fortress 2 – Classless Update". Valve. August 12, 2009. Retrieved May 21, 2009.

[22] "Gold Rush Update". *Team Fortress 2*. Valve Corporation. April 29, 2008. Retrieved May 1, 2008.

[23] "Team Fortress 2 – Sniper vs. Spy Update". Valve. May 13, 2009. Retrieved May 21, 2009.

[24] "Heavy Update: Arena Mode". Valve. August 18, 2008. Retrieved August 19, 2008.

[25] "Team Fortress 2 – Mann vs. Machine". Valve Corporation. August 13, 2012. Retrieved August 13, 2012.

[26] "Mann Up Mode FAQ". Valve Corporation. Retrieved August 16, 2012.

[27] Goldfarb, Andrew (October 26, 2012). "Team Fortress 2 Halloween Update Adds Zombies". IGN. Retrieved November 7, 2012.

[28] Peel, Jeremy (November 12, 2013). "Don't give up the ghost: Team Fortress 2 update enables Halloween all year round". PC Gamer. Retrieved November 12, 2013.

[29] "TF2 Official Blog: A Heavy Problem". *Team Fortress 2*. July 1, 2008. Retrieved September 28, 2008.

[30] "Meet the Scout". *Team Fortress 2*. Valve Corporation. Retrieved April 20, 2008.

[31] "Meet the Soldier". *Team Fortress 2*. Valve Corporation. Retrieved November 10, 2007.

[32] Goldstein, Hilary (May 23, 2007). "Team Fortress 2: Class Warfare". IGN. Retrieved September 21, 2007.

[33] "Pyro Update". *Team Fortress 2*. Valve Corporation. Retrieved June 18, 2008.

[34] "Meet the Demoman". *Team Fortress 2*. Valve Corporation. Retrieved November 10, 2007.

[35] "Meet the Heavy". *Team Fortress 2*. Valve Corporation. Retrieved November 10, 2007.

[36] "Meet the Engineer". *Team Fortress 2*. Valve Corporation. Retrieved November 10, 2007.

[37] Jungels, Jakob (July 3, 2008). "*TF2* Trading Cards – Part 2". *Team Fortress 2*. Valve Corporation. Retrieved July 5, 2008.

[38] "Meet the Sniper". *Team Fortress 2*. Valve Corporation. Retrieved June 17, 2008.

[39] "Seeds finalized for ESEA S12 TF2 Invite LAN". ESEA. October 8, 2012. Retrieved October 26, 2008.

[40] "ETF2L". Retrieved October 26, 2012.

[41] "UGC". Retrieved October 26, 2012.

[42] "AsiaFortress".

[43] Dransfield, Ian (April 28, 2015). "Matchmaking is coming to Team Fortress 2". *PC Gamer*. Retrieved September 9, 2015.

[44] Dunkin, Alan (June 1, 1998). "*Team Fortress* Full Speed Ahead". GameSpot. Retrieved June 12, 2006.

[45] "*Team Fortress Classic* (overview)". Planet *Half-Life*. Archived from the original on January 3, 2007. Retrieved December 2, 2006.

[46] "About *Team Fortress Classic*". PlanetFortress. Retrieved December 2, 2006.

[47] Dawson, Ed (November 11, 2000). "*Team Fortress 2* Q&A". GameSpot. Retrieved December 2, 2006.

[48] "Past Winners". GameCriticsAwards.com. Retrieved March 24, 2008.

[49] "*Team Fortress 2: Technology*". PlanetFortress. Retrieved April 5, 2007.

[50] Park, Andrew Seyoon (June 21, 2000). "New Engine for *Team Fortress 2*". *GameSpot*. Retrieved July 12, 2006.

[51] "Orange Box Interview". GameTrailers. August 29, 2007. Retrieved August 29, 2007.

[52] Berghammer, Billy (May 26, 2006). "Half-Life 2: Episode One Hands-On, Details, And Extensive Video Interview". Game Informer. Archived from the original on May 16, 2008. Retrieved May 5, 2008.

[53] "Friends 3.0 Pre-beta Interview". The Steam Review. January 26, 2006. Retrieved May 5, 2008.

[54] OnboardError (November 17, 2005). "HL2World's Interview With Doug Q+A". hl2world.com. Retrieved March 26, 2008.

[55] Berghammer, Billy (March 26, 2007). "The History Of *Team Fortress 2*". Game Informer. Archived from the original on April 3, 2007. Retrieved April 5, 2007.

[56] Valve Corporation (2007). *Team Fortress 2* **PC**. Level/area: In-game development commentary.

[57] Roper, Chris (July 14, 2006). "*Team Fortress 2* Teaser Impressions". IGN. Retrieved July 19, 2006.

[58] Ruymen, Jason (May 14, 2007). "Face-to-face with *Team Fortress 2* 's heavy". Valve Corporation. Retrieved May 5, 2008.

[59] Hellard, Paul (December 1, 2007). "Visual Design, Comic Game Action, with a purpose". CGSociety. Retrieved March 18, 2008.

[60] Reeves, Ben (March 12, 2010). "Writer's Block: Portal 2 Writers Roundtable". *Game Informer*. Retrieved March 13, 2010.

[61] Hatfield, Daemon (September 11, 2007). "*Team Fortress 2* Beta Begins Next Week". IGN. Retrieved May 5, 2008.

[62] McElroy, Justin (September 18, 2007). "*Team Fortress 2* beta now available". Joystiq. Retrieved May 5, 2008.

[63] Bokitch, Chris (September 18, 2007). "*Team Fortress 2* beta now open". Valve Corporation. Retrieved May 5, 2008.

[64] "Steam News Team Fortress 2 Blog Available". June 19, 2008. Retrieved June 21, 2008.

[65] "Steam announcement of updates (Meet the Sniper and Pyro unlockables)".

[66] Cherlin, Greg (April 2, 2009). "Wave goodbye to yer head, wanker". *Team Fortress 2*. Valve Corporation. Retrieved April 3, 2009.

[67] McDougall, Jaz (June 11, 2010). "Team Fortress 2 adds training mode, Mac support". *PC Gamer*. Retrieved January 15, 2011.

[68] Bailey, Kat (February 4, 2010). "Valve Adds Highlander Mode to Team Fortress 2". 1UP.com. Retrieved January 15, 2011.

[69] Chalk, Andy (December 21, 2010). "Valve Celebrates the Miracle of Australian Christmas". The Escapist. Retrieved January 15, 2011.

[70] Jackson, Mike (October 10, 2010). "Team Fortress Halloween update adds 'Headless Horsemann'". Computer & Video Games. Archived from the original on June 6, 2013. Retrieved August 13, 2012.

[71] Chalk, Andy (December 14, 2009). "Valve Brings Crafting to Team Fortress 2". The Escapist. Retrieved January 15, 2011.

[72] O'Conner, Alice (August 10, 2010). "Team Fortress 2 Getting Item Trading, Oodles of New Items Next Month". Shacknews. Retrieved January 15, 2011.

[73] Frushtick, Russ (September 30, 2010). "'Team Fortress 2' In-Game Store Launches, Here's How It Works". MTV. Retrieved January 15, 2011.

[74] Rose, Mike (May 6, 2011). "Team Fortress 2 Update Adds Replay Editor". Gamasutra. Retrieved May 6, 2011.

[75] "*Team Fortress 2* Update Released". Valve Corporation. March 18, 2010. Retrieved April 6, 2010.

[76] Team, TF2 (March 18, 2010). "Nice goin', pardner". *Team Fortress 2*. Valve Corporation. Retrieved April 6, 2010.

[77] Team, TF2 (January 13, 2010). "Yo, a little help here?". *Team Fortress 2*. Valve Corporation. Retrieved April 6, 2010.

[78] "*Team Fortress 2* Contribute". *Team Fortress 2*. Valve Corporation. April 2, 2009. Retrieved April 6, 2010.

[79] McWhertor, Michael (May 17, 2013). "Team Fortress 2's latest update, Robotic Boogaloo, is totally community-made". Polygon. Retrieved May 17, 2013.

[80] Nunneley, Stephany (June 12, 2013). "Team Fortress 2 community members have made $10 million". VG247. Retrieved June 12, 2013.

[81] Mejia, Ozzie (July 1, 2015). "Team Fortress 2 begins its three-month Gun Mettle Campaign tomorrow". Shack News. Retrieved July 2, 2015.

[82] Hillier, Brenna (July 2, 2015). "Team Fortress 2 Gun Mettle update delivers rare guns, weekly challenges from tomorrow". VG24/7. Retrieved July 2, 2015.

[83] Prescott, Shaun (July 2, 2015). "Team Fortress 2 update ushers in contracts, new maps and more". *PC Gamer*. Future plc. Retrieved July 2, 2015.

[84] Cox, Matt (October 6, 2015). "Team Fortress 2 Invasion update lands today". *PC Gamer*. Retrieved October 13, 2015.

[85] "*Team Fortress 2* 360 DLC Details Due 'Pretty Soon,' Fate of PlayStation 3 Content Uncertain". Shacknews. May 21, 2008. Retrieved May 21, 2008.

[86] "The Orange Box PS3 Patch Released". March 20, 2008. Retrieved December 23, 2008.

[87] Park, Andrew. "Team Fortress 2 Updated Hands-On Goldrush, New Achievements, New Items". GameSpot. Retrieved July 8, 2015.

[88] Loftus, Jack (March 5, 2008). "Valve wants free Team Fortress 2 expansions". GamePro. Archived from the original on March 8, 2008. Retrieved March 23, 2008.

[89] Faylor, Chris (August 22, 2008). "Valve Bringing Team Fortress 2 Updates to Xbox 360, Being Forced to Charge Gamers". Shacknews. Retrieved October 16, 2008.

[90] Faylor, Chris (March 3, 2010). "Valve Teases Announcement with Mystery Images; Steam Coming to Mac?". Shacknews. Retrieved January 15, 2011.

[91] "Wow, you guys are GOOD". Valve Corporation. June 10, 2010. Retrieved June 10, 2010.

[92] "Team Fortress 2 – The Mac Update! – Earbuds". Valve Corporation. June 2010. Retrieved September 30, 2010.

[93] "Steam for Linux Beta Now Available". Valve Corporation. November 6, 2012. Retrieved November 8, 2012.

[94] "Steam for Linux Beta Now Available to All". Valve Corporation. December 20, 2012. Retrieved January 13, 2013.

[95] Team, TF2 (February 14, 2013). "Team Fortress 2 is now on Linux". *Team Fortress 2*. Valve Corporation. Retrieved February 14, 2013.

[96] Gilbert, Ben (March 18, 2013). "Valve's Team Fortress 2 is Oculus Rift's first game, free 'VR Mode' update coming soon". Engadget. Retrieved March 18, 2013.

[97] Good, Owen. "Analyst Pegs Team Fortress 2 Hat Economy at $50 Million". *Kotaku*. Kotaku. Retrieved March 11, 2013.

[98] Crossley, Rob (June 23, 2011). "Valve: Team Fortress 2 is free forever". *Develop*. Retrieved June 23, 2011.

[99] Funk, John (June 20, 2011). "Valve Says It's Working on a Free-to-Play Game". The Escapist. Retrieved June 20, 2011.

[100] Miller, Patrick (March 7, 2012). "GDC 2012: How Valve made Team Fortress 2 free-to-play". Gamasutra. Retrieved August 24, 2012.

[101] Fahey, Mike (May 16, 2009). "Meet the Spy, Quite The Ladies Man". Kotaku. Retrieved May 17, 2009.

[102] Walker, Robin (May 18, 2009). "Getting to the bottom of things". *Team Fortress 2*. Valve Corporation. Retrieved May 20, 2009.

[103] Tolito, Stephan (May 31, 2009). "Valve Dreams Of Team Fortress 2 Movie, Divulges 'Meet The Team' Origins". Kotaku. Retrieved May 31, 2009.

[104] "It's Finally Time to Meet the Pyro". Kotaku.com. June 27, 2012. Retrieved November 9, 2012.

[105] "Meet the Truly Demented Mind of the Pyro". Gameverse. June 27, 2012. Retrieved November 9, 2012.

[106] Parkin, Simon (August 30, 2010). "Newell: Game-Makers Are Best Equipped To Turn Games Into Movies". Gamasutra. Retrieved August 30, 2010.

[107] http://kotaku.com/watch-the-newest-team-fortress-2-short-[5963177768]

[108] Daw, David (July 12, 2012). "Trying Out Valve's Movie Making Tools With the Source Filmmaker". Retrieved July 13, 2012.

[109] O'Conner, Alice (August 14, 2009). "Valve Talks Team Fortress 2 Comic Book Plans, Movie and TV Show Possibilities". Shacknews. Retrieved July 8, 2010.

[110] Chalk, Andy (July 5, 2010). "TF2 Update: The Engineer With the Golden Wrench". The Escapist. Retrieved July 8, 2010.

[111] Funk, John (December 9, 2009). "Valve Teases TF2 Demoman and Soldier Updates". The Escapist. Retrieved July 8, 2010.

[112] Rose, Mike (July 11, 2011). "Comic Book Based On Valve Strips Coming This November". Gamasutra. Retrieved July 11, 2011.

[113] Good, Owen (August 12, 2012). "Team Fortress 2's Gray Mann Surfaces as Signs Point to All-Robot Faction". Kotaku. Retrieved August 13, 2012.

[114] Hillier, Brenna (August 12, 2012). "Team Fortress 2 to add third, robotic faction – rumour". VG247. Retrieved August 13, 2012.

[115] "Valve Interview Part 2: *Left 4 Dead* Demo Potential, the Evolution of Steam, and More". Shacknews. May 23, 2008. Retrieved July 2, 2008.

[116] "Pre-Order Giveaway Madness!". Valve Corporation. November 2, 2009. Retrieved September 30, 2010.

[117] "Free Hats!". Valve Corporation. October 6, 2010. Retrieved October 6, 2010.

[118] Francis, Tom (August 20, 2010). "Valve on the future of Team Fortress 2. Part Two". PC World. Retrieved August 20, 2010.

[119] Sonic & All-Stars Racing Transformed – PC Trailer on YouTube

[120] Hamilton, Kirk (June 3, 2013). "Team Fortress 2 Gets Its Very Own TV Commercial". Kotaku. Retrieved June 3, 2013.

[121] "*Team Fortress 2* Reviews". GameRankings. Retrieved September 1, 2008.

[122] "*Team Fortress 2* (PC: 2007): Reviews". Metacritic. Retrieved September 1, 2008.

[123] Elliot, Shawn (October 10, 2007). "*Team Fortress 2* PC Review". 1UP.com. Retrieved May 2, 2008.

[124] Bradwell, Tom (October 10, 2007). "*Team Fortress 2* Review". Eurogamer. Retrieved May 2, 2008.

[125] Watters, Chris (May 3, 2008). "*Team Fortress 2* for PC review". GameSpot. Retrieved July 8, 2015.

[126] Accardo, Sal (October 10, 2007). "*Team Fortress 2* Review". GameSpy. Retrieved May 2, 2008.

[127] "*Team Fortress 2* Review". GamesRadar. Retrieved April 24, 2015.

[128] Burt, Andy (October 10, 2007). "*The Orange Box* Review". GamePro. Archived from the original on October 11, 2007. Retrieved May 2, 2008.

[129] "GameSpy's Game of the Year 2007: *Team Fortress 2*". GameSpy. Archived from the original on December 31, 2007. Retrieved December 22, 2007.

[130] "2007 1UP Network Editorial Awards". 1UP.com. Retrieved February 18, 2008.

[131] "GameSpy Game of the Year 2007: Multiplayer". GameSpy. Archived from the original on January 8, 2008. Retrieved April 6, 2008.

[132] "GameSpy's Game of the Year 2007: Special Awards". GameSpy. Archived from the original on January 4, 2008. Retrieved April 6, 2008.

[133] "IGN Best of 2007: PC Best Artistic Design". IGN. Archived from the original on November 17, 2013. Retrieved February 18, 2008.

[134] "11th Annual Interactive Achievement Awards". AIAS. Archived from the original on March 19, 2008. Retrieved April 24, 2008.

[135] "Spike TV Announces 2007 'Video Game Awards' Winners". Spike TV. December 7, 2007. Retrieved July 8, 2015.

[136] Booth, Mike (December 21, 2009). "Erectin a dispenser". Valve Corporation. Retrieved January 22, 2010.

2.47.6   External links

- Official website

- The *Team Fortress 2* page at the official site of *The Orange Box*

- Official *Team Fortress* Wiki

2.48   They (video game)

They (stylized as *THEY*) is a first-person shooter that was formerly under development by Metropolis Software. When the studio was purchased by CD Projekt, development was halted in order to focus on *The Witcher 2: Assassins of Kings*. According to CD Projekt RED co-founder Marcin Iwinski the game is not completely cancelled, stating "...we have by far not buried *They*, and we would really like to return to it." [1] However, on June 21, 2015, CD Project RED has stated that the project "was cancelled many years ago." [2]

In-game screenshot from Games Convention 2007

2.48.1   Gameplay

The single player campaign is up to twelve hours long with a story driven setting. Multiplayer consists of deathmatch and team deathmatch missions.[3]

One of the distinctive elements of the game is the option to modify player's weapon with the components they collect. This feature, known as the "Weapons Tuning System", allows players to adjust their weaponry for the role they will face in the future.

The demonstration video revealed that some of the machines are controlled by aliens known as Phantoms, who give them a greater level of initiative and intelligence. The Phantom possessed machines are easily distinguishable from ordinary machines by their red-colored form.

2.48.2   Plot

Set in the year 2012, players will take the role of a soldier in England, who works with his comrade, a pilot. Terrorism has become a major problem; however, that changes when mysterious robots attack. The machines' intentions are unknown but the player soon discovers aliens called Phantoms are controlling them.

2.48.3   References

[1] Purchese, Robert (29 January 2010). "CD Projekt suspends FPS They". Eurogamer. Retrieved 10 February 2010.

[2] "The project was cancelled many years ago.". Twitter. 2015-06-21. Retrieved 2015-06-22.

[3] "The game will feature a story driven single player mode with many different characters to speak and act with – it will have a playtime of around 20 hours and Unique Weapon system".

2.49 This Is Vegas

This Is Vegas is an unreleased video game that was in development at Surreal Software and was to be published by Warner Bros. Interactive Entertainment on Microsoft Windows, the PlayStation 3, and the Xbox 360. In 2010 development on the game was reportedly cancelled.*[1]

2.49.1 Development

In January 2006 IGN reported that game development studio Surreal Software had posted job listings for artists to work on "an upcoming PlayStation 3 and Xbox 360 game that is, as of yet, without a title." *[2] In February 2008 the publishing company and parent of Surreal Software, Midway Games, announced that *This Is Vegas* was in development at the studio. Midway released several screenshots and stated that it planned to release the game on Microsoft Windows, the PlayStation 3, and the Xbox 360 in the fourth quarter of 2008.*[3] That same month Midway released the debut trailer for the game.*[4]

During Midway's bankruptcy proceedings in 2009, the company sold the majority of its assets to Warner Bros. Interactive Entertainment, including *This Is Vegas* and developer Surreal Software.*[5] In August 2010, *Computer and Video Games* reported that Warner Bros. cancelled development on the game.*[1] Midway and Warner Bros. spent between $40 million and $50 million developing the game.*[6]

2.49.2 References

[1] Ingham, Tim (2010-08-24). "Is this gaming's biggest ever waste of money?". *Computer and Video Games*. Retrieved 2013-12-14.

[2] Dunham, Jeremy (2006-01-19). "Surreal Preps for Next-Gen". IGN. Retrieved 2013-12-15.

[3] Kietzmann, Ludwig (2008-02-06). "Midway introduces 'This is Vegas' to Xbox 360, PS3, PC". Joystiq. Retrieved 2013-12-15.

[4] Garratt, Patrick (2008-02-08). "This is Vegas – first trailer". vg247. Retrieved 2013-12-15.

[5] McWhertor, Michael (2009-07-10). "Warner Bros. Now Owns Midway, Mortal Kombat". Kotaku. Retrieved 2013-12-14.

[6] Van Zelfden, N. Evan (2009-02-16). "What's Killing the Video-Game Business?". *Slate*. Retrieved 2013-12-14.

2.49.3 External links

- Official website
- Debut trailer
- Trailers and videos at GameTrailers

2.50 Tiny Toon Adventures: Defenders of the Universe

Tiny Toon Adventures: Defenders of the Universe (also known as *Tiny Toon Adventures: Defenders of the Looni-verse*[1]) is the name of a Tiny Toon Adventures video game initially scheduled for release in 2001, but was eventually cancelled for unknown reasons. It was developed by a Japanese video game studio, Treasure and it was originally slated for the PlayStation 2.

2.50.1 Plot

On a planet made almost entirely out of gold located in the Jewel system, live the Bullions. They have lived there peacefully for many years, until their home was attacked by the Drizzletroopers. Two of them manage to escape, Zig and Zag, to the planet Earth. The Leader of the Drizzletroopers learns of this and has someone go after the two. As morning comes by, Buster Bunny and the gang attend school. However, a strange robot, disguised as Bugs Bunny, appears in their class rather than their teacher.

When his disguise is blown, the group chase after the fleeing robot. After stopping the robot from getting away, he makes an attempt to kill everyone, but ultimately fails, leaving a hole in the ground. There, Zig and Zag appear out of the hole, startling the group. The strange creatures explain their situation and the group agrees to help them.

However, Shirley appears telling the group a great danger will fall upon the earth if they help them. Despite her warning, the group decides to help the creatures out. They all enter the hole into the ground, leading them into yet another new adventure...

2.50.2 Cast

Nearly all of the surviving members of the original main cast of *Tiny Toon Adventures* reunited to provide the characters for the game (with Charlie Adler returning as the voice of Buster and Billy West replacing the late Don Messick as the voice of Hamton).

- Charlie Adler - Buster Bunny, Zig

- Tress MacNeille - Babs Bunny, Zag

- Joe Alaskey - Plucky Duck

- Billy West - Hamton Pig, Maximus Flush

- Danny Cooksey - Montana Max, Evil Montana Max

- Gail Matthius - Shirley the Loon

- Kath Soucie - Fifi La Fume

- Cree Summer - Elmyra

- Maurice LaMarche - Dizzy Devil

- Frank Welker - Gogo Dodo, Furrball

2.50.3   Development termination

The game's original official website, www.dotu.net, no longer exists. Tetsuhiko Kikuchi *(aka Han)*, the director of the game, has hinted on his homepage *(Japanese)* that the game "will forever go unreleased."

An article on IGN, written back in 2004, had suggested that the game had always been intended for release.*[2] It said that at the end of the article that the game was "still officially slated to come out sometime (not necessarily in 2004)". However, years later, nothing new has been mentioned about the game. With the releases of the next generation consoles, Xbox 360, PlayStation 3 and Wii, the game's release is very unlikely. The main page at IGN now has *canceled* for the US release date.*[3]

Exact reasons for the game's cancellation remain unknown, but are generally believed to be unrelated to the state of the product itself. Publisher Conspiracy Entertainment was in dire financial shape at the time and several other projects of theirs quietly disappeared during this same time. They cited unspecified "business complications" for the extended hold up. Their previous partnership with Treasure, Tiny Toons: Scary Dreams was manufactured in 2002, but its existence could not be confirmed until it appeared on eBay in 2005. Defenders of the Universe was evaluated by the ESRB, which some take as evidence that the game was in a near-final state.

2.50.4   Prototype

On February 25, 2009, an image of a prototype of the game was released by a member of the internet forum, Lost Levels. The prototype confirmed many suspicions about the game, including its advanced state of completion, and the involvement of Rakugaki Showtime director Tetsuhiko Kikuchi. The title appears feature-complete, apart from a few errors in the in-game text.

2.50.5   References

[1]   "Tiny Toons Adventures: Defenders of the Looniverse". IGN. May 13, 2001.

[2]   Douglass C. Perry (2002-05-30). "Missing in Action: The Lost Games of the PlayStation 2, Part I - PS2 Feature at IGN". Ps2.ign.com. Retrieved 2012-02-16.

[3]   "Tiny Toon Adventures: Defenders of the Universe - PS2 - IGN". Ps2.ign.com. Retrieved 2012-02-16.

2.50.6   External links

- Official Developer's Website (*requires Japanese support*)

- The World of Tiny Toon Adventures Game Information Page

2.51   Ugo Volt

Ugo Volt (formerly ***Prospects of Mayhem***) is a game by Rogério Varela, set in the 22nd century, in a post-apocalyptic Lisbon, almost destroyed by global warming. It was also the first Portuguese game to be featured at E3 2006.*[2]

With no news from the game since 2007, with the website now under control of a third-party, and with the official website of the company non-operational for some time (as of 2010), it is assumed the game was either cancelled or is not in active development anymore.

2.51.1   Gallery

- Ingame screenshot

- Ingame screenshot

2.51.2   References

[1]   "Ugo Volt". Retrieved February 23, 2011.

[2]   "PRIMEIRA EMPRESA PORTUGUESA A EXPÔR NA E3" (PDF) (Press release) (in Portuguese). Move Interactive. 2006-05-05. Retrieved 2007-12-26.

2.51.3   External links

- *Ugo Volt* Official website (planned website, now under control of a third-party)

- *Move Interactive* official website

- News and Videos about *Ugo Volt* (Portuguese)

- Trailer at IGN

- Videos (also in HD) at GameTrailers.com

2.52 Warcraft Adventures: Lord of the Clans

This article is about the unreleased computer game. For the book based on the same story, see Warcraft: Lord of the Clans.

Warcraft Adventures: Lord of the Clans was a black comedy point-and-click adventure video game under development by Blizzard Entertainment, set in the *Warcraft* universe, and cancelled before its release. American company *Animation Magic**[1] was out-sourced due to their experience in classical two-dimensional animation to produce the twenty-two minutes of fully animated sequences, the game's artwork, the coding of the engine and the implementation of the sound effects. Blizzard provided all the designs, the world backgrounds, sound recording and ensured storyline continuity. Four or five months after Blizzard had released Battle.net and *Warcraft II: Beyond the Dark Portal* had shipped, Blizzard began development on *Lord of the Clans*, that would be cancelled just over a year later.

2.52.1 Development and cancellation

Warcraft Adventures: Lord of the Clans was originally slated for a fourth-quarter 1997 release; however it was pushed back until the end of 1998. This was a result of unforeseen technical problems coupled with communication limitations between Blizzard and the Russian animators at Animation Magic. The game had been in development for over a year: nearly all features, puzzles, and areas were in place, the voice acting had been recorded, and much of the animation was complete, yet Blizzard was not confident with their title. Blizzard hired Steve Meretzky, creator of *A Mind Forever Voyaging* and *The Hitchhiker's Guide to the Galaxy* video games, as a design specialist to help refine the puzzles and make them further cohesive with the narrative. Meretzky spent two weeks with the developers looking over the game for up to fourteen hours a day and it was decided that sequences of the game had to be rewritten which would involve more animation and more dubbing.

However, as the 1998 *Electronic Entertainment Expo* (E3) in Atlanta was approaching, Blizzard became increasingly aware that implementing the proposed changes would result in them being unable to meet their already extended 1998

deadline. LucasArts had released their title *The Curse of Monkey Island* in the fall of 1997, and had announced their next adventure game title *Grim Fandango* sporting a 3D engine. In comparison, producer Bill Roper felt *WarCraft Adventures* looked dated;

I think that one of the big problems with *Warcraft Adventures* was that we were actually creating a traditional adventure game, and what people expected from an adventure game, and very honestly what we expected from an adventure game, changed over the course of the project. And when we got to the point where we cancelled it, it was just because we looked at where we were and said, you know, this would have been great three years ago.

After over a year of hard work, press tours, magazine covers, and fan fervor Blizzard announced that *Warcraft Adventures: Lord of the Clans* was cancelled days before E3. Within hours of the announcement fans of the series formed an online petition, demanding the project be resurrected. On May 22, 1998, Blizzard responded via their website;

Blizzard Announcement 22 May 1998
Press Desk: Blizzard Cancels *Warcraft Adventures*

Blizzard wants to take a minute to respond to the *Warcraft Adventures* petition that is circulating on the Internet. First, we want to express our gratitude to the Warcraft fans that took the time to organize such an effort. We recognize that the cancellation of *Warcraft Adventures* has disappointed some of our customers, and we appreciate that they have shared their opinions with us.

Secondly, we want let you know that stopping development was not a decision that was taken lightly. It was a hard call to make, but each of us knows that it was the right choice. The cancellation was not a business or marketing decision or even a statement about the adventure genre. The decision centered around the level of value that we want to give our customers. In essence, it was a case of stepping up and really proving to ourselves and gamers that we will not sell out on the quality of our games.

And finally, we hope that Warcraft fans will consider our track record and trust our judgement on ending the project. The cancellation of *Warcraft Adventures* does not signal the demise of Azeroth. We have every intention of returning to the *Warcraft* world because there are still chapters to be told. We will keep you informed as we announce future Warcraft plans.

Despite their press release, rumours still persist the game was cancelled due to projected low sales from the deteriorating market of the adventure game genre. Even though the game was cancelled, Blizzard felt the story itself too important to ignore and hired an author to adapt it into a novel. The author contracted to scribe it was unable to complete the book on time, so *Star Trek* novelist Christie Golden was hired to write the novelization based on scripts and outlines provided by *Warcraft* universe co-creator, Chris Metzen, and had to be completed within six weeks. The book was released under the title *Warcraft: Lord of the Clans* by Pocket Books and is considered canonical. *Warcraft: Lord of the Clans* is the second novel based in the *Warcraft* universe.

Blizzard returned to the Azeroth setting in 2002, with the release of *Warcraft III: Reign of Chaos*. Even though the game exploring his storyline had been cancelled, Thrall played a major role in *Warcraft III* and the subsequent massively multiplayer online role-playing game *World of Warcraft* and his story as outlined in the novel is considered canon, according to the *Warcraft III* manual's backstory.

In March 2010, a video of 20 minutes of gameplay was uploaded to a Russian gaming website, proving that someone outside of Blizzard Entertainment possesses a version of the game.[*][2] The video is also available on YouTube (where it is labeled as from an alpha version of the game).[*][3] In February 2011 a series of 11 gameplay videos (labeled as from a beta version of the game) were uploaded to YouTube (in December 2014 and January 2015 the 12th and 13th videos respectively have been added) walking through entire available gameplay.[*][4]

2.52.2   Characters

Thrall was voiced by Clancy Brown. Orgrim Doomhammer was voiced by Peter Cullen and Drek'thar (the shaman) was voiced by Tony Jay. Additional voices were done by Bill Roper[*] and Zul'jin (Troll leader), Nazgrel (the shaman), Gazlowe, Durotan, Rend, Maim, Kargath Bladefist, Deathwing were voiced by unknown voice actors.

2.52.3   Notes

^ Bill Roper produced the voices for all the characters in *Warcraft: Orcs & Humans* and many of the voices in *Warcraft II: Tides of Darkness*, however as primarily a producer he is not a union actor so was not allowed to be used for *Warcraft Adventures*. However they could legally use his previously recorded work sparingly throughout the game.

2.52.4   See also

- *StarCraft: Ghost* - another canceled Blizzard title

2.52.5   References

Footnotes

[1] Hardcore Gaming 101: Zelda: Wand of Gamelon / Link: Faces of Evil

[2] http://forums.ag.ru/?board=ag_files&action=display&num=1268337910&start=0

[3] http://www.youtube.com/watch?v=ccqJ9W_E3jI

[4] http://www.youtube.com/user/0manbiker0

Notations

- *Warcraft Adventures: Lord of the Clans* feature by GameSpot

- Interview with author of *Lord of the Clans*, Christie Golden

- *Art of Warcraft*, Brady Games, Pearson Education, Indianapolis, Indiana, 46290, United States of America, 2002. ISBN 0-7440-0081-5

2.52.6   External links

- Warcraft Adventures: Lord of the Clans on Wowpedia, a *Warcraft* wiki

Chapter 3

Text and image sources, contributors, and licenses

3.1 Text

- **Vaporware** *Source:* https://en.wikipedia.org/wiki/Vaporware?oldid=679930767 *Contributors:* Damian Yerrick, Bryan Derksen, Taw, Fubar Obfusco, Roadrunner, Michael Hardy, Nixdorf, Tompagenet, Ixfd64, Tregoweth, Angela, Nanobug, Netsnipe, Grin, WhisperToMe, Furrykef, Populus, Bloodshedder, Dpbsmith, Rhsatrhs, Robbot, Paranoid, Benwing, RedWolf, Sbisolo, ZimZalaBim, Samrolken, Postdlf, Lord Bob, DHN, Kzhr, Unfree, Alexwcovington, Tom harrison, Bkonrad, Gracefool, Zoney, Gregb, Macrakis, Golbez, WikiFan04, TerokNor, Andycjp, Slowking Man, Sonjaaa, Ihavenolife, Jh51681, Louisisthebest 007, Chmod007, Mjuarez, Rich Farmbrough, ObsessiveMathsFreak, YUL89YYZ, Weyoun6, Bender235, Ylee, Kaszeta, CanisRufus, Tverbeek, Erath, 23skidoo, Marblespire, Robotje, Spug, Scott Ritchie, Bob rulz, Ghostalker, Moanzhu, Ashley Pomeroy, ChaosFish, Titanium Dragon, Scott5114, 2mcm, Sleigh, Cédric, Zntrip, Defixio, Lkinkade, Daveydweeb, Richard Arthur Norton (1958-), Mindmatrix, Tabletop, Hbdragon88, Cbdorsett, Hotshot977, GregorB, Marudubshinki, Kesla, Pastel-ink, Magister Mathematicae, Rjwilmsi, Attitude2000, CDCAA18D, Ligulem, KarlFrei, Ysangkok, Itinerant1, Gurch, Mitsukai, Intgr, D.brodale, Chobot, Eric B, George Leung, Badanagram, Schmancy47, RobotE, Beltz, Hede2000, Bhny, AstrixZero, Manop, Akhristov, Mipadi, Daemon8666, Thalter, Brendanurbanwarrior, Zwobot, Epa101, DeadEyeArrow, Ms2ger, Mugunth Kumar, 21655, CWenger, Snaxe920, Olav Bringedal, Saikiri, One, NetRolller 3D, A bit iffy, SmackBot, Jillcvm, SergeantPepper, IstvanWolf, Kmarinas86, Andyzweb, CKA3KA, Donbas, Happylobster, Miquonranger03, Mdwh, The Moose, Nintendude, Emurphy42, KieferSkunk, SuperDuffMan, Addshore, Hateless, Warren, Hgilbert, ShadowUltra, Alsh, Gbr~enwiki, ModemManCJ, Ohconfucius, Esrever, JackLumber, Eridani, Genisock2, Romeu, Hu12, JoeBot, Blakegripling ph, Tawkerbot2, Elekas, Godgundam10, Insolectual, FatalError, AdamJohnso, CmdrObot, Wikipedian06, Ibadibam, Penbat, Drewdowling, Argus fin, Kafka1251, Davidrossiter, Xenelia, Kozuch, Lordhatrus, JAF1970, Lord Hawk, Tcp 200, Qwyrxian, Sagaciousuk, Wlog90, X201, Spud Gun, AntiVandalBot, Nshuks7, Erfa, Vendettax, Spartaz, Tws45, Klow, Arsenikk, NapoliRoma, TostitosAreGross, Nabkawe, BCube, Kerotan, Steveprutz, Magioladitis, Bongwarrior, Cadsuane Melaidhrin, Hzoi, JaffaCakeLover, SmashTheState, LindsieandLance, Richiekim, Tidytibs, Dispenser, Ethd, Unflavoured, BigHairRef, Flatscan, Chaos5023, Elbbom, Yf metro, Liko81, Valaqil, Bob f it, Feudonym, Mantisia, Winnd, Mikemoral, Psbsub, SheepNotGoats, Oldag07, Fraser J Allison, RJaguar3, Flyer22 Reborn, Juandope, Loren.wilton, Martarius, ClueBot, Foxj, Sebquantic, Trivialist, Toad of Steel, Jack Bauer7, John Nevard, Archon Shiva, Sun Creator, Sin Harvest, A plague of rainbows, Tezero, Stevenrasnick, DumZiBoT, Mjharrison, Feinoha, Insanekaosx, ErkinBatu, Dekart, JCDenton2052, ESO Fan, Cyclonebiskit, Dsimic, Rustyjames 2000, Addbot, Drakkenmensch, Fyrael, Ironholds, Jbh001, TheFreeloader, Jarble, Yobot, KgKris, Pcap, Jabberwockgee, GamerPro64, AnomieBOT, 威 因, Carlsotr, LilHelpa, Xqbot, Masterius, SassoBot, TheIncredibleNix, FrescoBot, Allengittelson, Voronwe, HamburgerRadio, Citation bot 1, Jonesey95, Nigtv, Dwigs, Jonkerz, Airbag190, RjwilmsiBot, Zaqq, Akjar13, GoingBatty, RenamedUser01302013, Evanh2008, Informer3X, Forsak3n 1, Lacon432, H3llBot, Captain Screebo, Someguy432, SporkBot, Radixen, Johnny421, Ego White Tray, ClamDip, MuzRat, ClueBot NG, Pantergraph, Matthiaspaul, Rhain1999, Zeko927, Goodyob, Shovan Luessi, Helpful Pixie Bot, 254Jackson, Curb Chain, Awsometilthegrave, Nexusmason, Bockrocker2, IluvatarBot, Tommyncfc, BattyBot, Morc66, Cyberbot II, ChrisGualtieri, TheJJJunk, Spz0, Mogism, Makecat-bot, Adeshina365, Strawb309, CokeHanx, Monkbot, Zaddikskysong, Twist295 and Anonymous: 454

- **List of vaporware** *Source:* https://en.wikipedia.org/wiki/List_of_vaporware?oldid=682097423 *Contributors:* Zundark, Frecklefoot, Dogface, JonathanDP81, Bearcat, Altenmann, JesseW, DO'Neil, AlistairMcMillan, Kmweber, TreyHarris, Urvabara, Discospinster, Lovelac7, Bo Lindbergh, Kenb215, Richard W.M. Jones, 23skidoo, John Vandenberg, GatesPlusPlus, Astrophizz, Debolaz, Gargaj, Colin Kimbrell, Evil Monkey, Jakes18, Red dwarf, Daveydweeb, Richard Arthur Norton (1958-), Firsfron, LeonWhite, GregorB, BD2412, Rjwilmsi, W3bbo, Intgr, DrIdiot, Tyro, Antilived, Pigman, Brooza, Malcolma, HQ, Ycherk03, IsUsername, 2fort5r, JLaTondre, Curpsbot-unicodify, Mardus, Dandelions, NetRolller 3D, Elomis, A bit iffy, SmackBot, CitrusC, Nsayer, Nil Einne, Cyrax777, Marc Kupper, Happylobster, Mdwh, Xyzzyplugh, Warren, Tehw1k1, S@bre, Ourai, Tarcieri, Artman40, Babak 91, TJ Spyke, Cnbrb, Crkbbyx, Danial79, Loopkid, Iced Kola, GargoyleMT, El Ucca, AndrewHowse, Petemwah, Dancter, Wermlandsdata, Renamed user 1752, ItKatyPerry, TexasFury, Tut21, Erfa, FireDemon~enwiki, JAnDbot, Furyhunter600, NapoliRoma, Davidjk, Bronek~enwiki, Richiekim, Guticb, DAOWAce, NecroSen, Group29, Aun'va, Flatscan, GCFreak2, 87speedos, Stateofshock, Vashner, Bzdenok, Steven J. Anderson, TheChode, Zlnetworks, NATO.Caliber, Jerryobject, Lightmouse, Sliwers, KingCapybara, Martiian000, Soilworker~enwiki, Iulian28ti, Sebquantic, Frmorrison, Mushinronjya, Kurosei, Tr6c, Mmorg, JCDenton2052, Jarble, Arbitrarily0, PlankBot, Tohd8BohaithuGh1, AnomieBOT, Exp HP, Randomnessness, Jim1138, C00LsPoT GP02, Richard BB, Logboekverhaal, Silesianus, Klubbit, Dwigs, Grossa144, Pimmfritt, Dewritech, Koldcuts, Ego White Tray, Rhain1999, Nathan2055, BattyBot,

Khazar2, Shivertimbers433, Mogism, PingasTAM, Myconix, Hayesinator, Skr15081997, Blart versenwald, Traderdata, Gamingforfun365, Kon-Newton, Grizzie217 and Anonymous: 237

- **Amiga Walker** *Source:* https://en.wikipedia.org/wiki/Amiga_Walker?oldid=634105432 *Contributors:* RussBot, Harryboyles, Sambot, Eivind F Øyangen, Horologium, Marko75, Keeper Garrett, Sfan00 IMG, XLinkBot, Darko Sola and Anonymous: 7

- **Asus Transformer Book Duet** *Source:* https://en.wikipedia.org/wiki/Asus_Transformer_Book_Duet?oldid=683311130 *Contributors:* Smyth, Wavelength, ViperSnake151, Cydebot, Keith D, CommonsDelinker, Admiralthrawn 1, Andy Dingley, Aepko, TutterMouse, AnomieBOT, Ruby2010, LittleWink, Dewritech, Newyorkadam, BG19bot, VanishedUser 2313214sad1, Jodosma, Ethically Yours, Some Gadget Geek and Anonymous: 7

- **Beyond Good & Evil 2** *Source:* https://en.wikipedia.org/wiki/Beyond_Good_%26_Evil_2?oldid=685504234 *Contributors:* Topbanana, Tgies, Alansohn, Gary, CyberSkull, Godheval, RJFJR, Zxcvbnm, Ground Zero, Koveras, RJC, Pwlodi, Tony1, Occono, Falcon9x5, Dwb122, Rehevkor, Hmains, Cbebop007, Masem, SkyWalker, SuperTank17, Cydebot, Soetermans, Thijs!bot, X201, Redsparta, KitAlexHarrison, JNW, Giggy, LoganTheGeshrat, CommonsDelinker, Anonymous Keeper of Records, Svetovid, VolkovBot, LeilaniLad, Beem2, JayC, Klatys~enwiki, Magiske~enwiki, Djistus, Fratrep, Svick, TaerkastUA, WikiLaurent, Wonchop, ImageRemovalBot, Martarius, ClueBot, Uncle Milty, Samurai Cerberus, TapDatApp, XLinkBot, Addbot, Cxz111, Megata Sanshiro, G0T0, LaaknorBot, Contributor777, Yobot, Ptbotgourou, Mr T (Based), AnomieBOT, BlazerKnight, Garcon0101, Wrelwser43, Fran Heunther, Starkos, Finalius, ProtoDrake, Tommiej, DrilBot, Porcho, Lightlowemon, Ayken, Johnyt123, HandyBurb, Bilbo571, Lord Psyko Jo, ClueBot NG, Flax5, Sni56996, SNAAAAKE!!, JS-Tactics, Windows.dll, Anijatsu, Videogamer2000, Iseesky, Celeski96, AdrianGamer and Anonymous: 124

- **Black Mesa (video game)** *Source:* https://en.wikipedia.org/wiki/Black_Mesa_(video_game)?oldid=687955726 *Contributors:* Wwwwolf, Ixfd64, Karada, Daniel Quinlan, K1Bond007, Robbot, Chocolateboy, Caknuck, Asparagus, OverlordQ, Gscshoyru, Ehamberg, Asqueella, Johan Elisson, Discospinster, Jovrtn, Thunderbrand, Billymac00, Smalljim, Firefox~enwiki, Alansohn, CyberSkull, Jtalledo, Plumbago, Riana, RJFJR, New Age Retro Hippie, Marasmusine, LOL, RabidMonkey, Rjwilmsi, Linuxbeak, Nick R, Yar Kramer, Winhunter, RexNL, DrIdiot, Chobot, Raider Duck, YurikBot, I need a name, Hede2000, NateDan, PatCheng, Outlive, Retired username, Brandon, Aaron Schulz, Richardcavell, ReCover, DaltinWentsworth, WormNut, Mad onion, Some guy, Kingboyk, Rehevkor, Tom Morris, Gundam Bass, SmackBot, Reedy, Ma8thew, C.Fred, Chairman S., David Fuchs, Xaosflux, Bluebot, SynergyBlades, Can't sleep, clown will eat me, Gohst, Anoriega, DMacks, Kotjze, S@bre, LeoNomis, GameKeeper, Midkay, MutantBeef, Pizzahut2, KLLvr283, Sam Kellett, Teancum, BMSKalashnikov, CaptainVindaloo, Wes less, Bluseychris, Mr Stephen, Kizzatp, Tyhopho, Delta759, Masem, HelloAnnyong, Bbenjoe, DougClayton4231, Lukeydukey, LearningKnight, Protiek, SkyWalker, Daedalus969, JForget, CmdrObot, Zarex, Addict 2006, Mika1h, Tim1988, Cydebot, Xkesterx, Besieged, Mato, Odie5533, Tawkerbot4, DumbBOT, Mooseofshadows, Omicronpersei8, BetacommandBot, Raminator~enwiki, Unicyclopedia, Hervegirod, Gamer007, Mojo Hand, Headbomb, X201, Grayshi, Redgrassbridge, F l a n k e r, Morginq, AntiVandalBot, Robzz, Snake-4c-merc, Erfa, Pixelface, Mizery Made, Hewinsj, ProjectPlatinum, Acroterion, Geniac, MMorris192, Bongwarrior, CTF83!, JeffJonez, Pkaulf, Godalmighty83, T e r o, Varsindarkin, Katana314, Knos, David Munch, J.delanoy, Trusilver, El Carnemago, Darth Mike, Arite, GRiM-reapa, QuasiAbstract, Rolod~enwiki, Raw bean, STBotD, Skinnattittar, Billwaa, JavierMC, Gracz54, Jeff G., Budtard, JeanPaul12, Beardsall, Collision, Martin451, Megasquid500, NHRHS2010, Undead warrior, MuzikJunky, Jenn0 Bing, Caltas, RJaguar3, Flyer22 Reborn, Radon210, Tenyu huang, Jscorp, MikeZuniga, Troy 07, Aragorn12712, Martarius, ClueBot, The Thing That Should Not Be, Sebquantic, ElectricalTill, Alexandersedov, Mezigue, Ms 2007 Cn, Dylan620, Mr. Someguy, Cirt, PMDrive1061, Paulcmnt, Beastmonkey, T3hphoenix, Coozins, FrostedBitesCereal, Zarnivop, XLinkBot, Noskap, Addbot, Willking1979, Ronhjones, Leszek Jańczuk, Doniago, Tide rolls, MuZemike, Yobot, Ptbotgourou, Senator Palpatine, Fraggle81, TaBOT-zerem, PerfectFaro, Angsc09, TheBigJagielka, AnomieBOT, Wasabi01, Bmsreader, JackieBot, Bluerasberry, Subzero-Mick, Brightgalrs, Xqbot, Dunnybrusher, CoolingGibbon, Jeffrey Mall, Jmundo, Deni42, GVilKa, SchnitzelMannGreek, Sesu Prime, Fortdj33, Wikipe-tan, Kausill, JMS Old Al, Aldy, Zachayes, DrilBot, Hollow334, Ladydarkside1, Zerosandz, Notedgrant, Elvis Thong, NoRicov, Jessicagill75, TagsBot, Xzoop, Jackoffison, HighFived, Barras, TrollMan69, Raminatorz, Veitschwul, JAWNFREEDMAN, GAAAARY, 1234cuminmymouth, CAREBAREHUBI, Mango SENTINEL, Hl2dq, Lammidhania, Reach Out to the Truth, Derild4921, Onel5969, RjwilmsiBot, Saftorangen, Slon02, Jroswen, JimBonerston, EmausBot, Immunize, GoingBatty, Slightsmile, Wikipelli, ZéroBot, Chrisgooverthere, Webmanners, Bilbo571, H3llBot, Boatscaptain, Chrisisawesome44, Donner60, MrConti, ChuispastonBot, Datchan, AnddoX, ClueBot NG, Shaddim, Metathink, Halokid12, Jwchong, Flynn58, Thekken2, Player017, Chriscshunter, Lunferd, The1337gamer, BattyBot, Seth.mccarus, Rockguy 91, Dissident93, Makecat-bot, Cman2k, WikiTransfer, GabeIglesia, Nicereddy, The Anonymouse, Rupert loup, F6Zman, John33213, Dustin V. S., DavidLeighEllis, Wowextwo2, Qed237, G S Palmer, Leftshift, Vieque, JamesMort2011, Hakken, Joshhubi, RoadWarrior445 and Anonymous: 444

- **Bob's Game** *Source:* https://en.wikipedia.org/wiki/Bob'{}s_Game?oldid=669322498 *Contributors:* Twilsonb, EldKatt, Sn0wflake, Axeman89, Tsuba~enwiki, Foltor, Grafen, Nikkimaria, SmackBot, Sam mishra, Axem Titanium, Teancum, Kingoomieiii, Mika1h, X201, Lordmetroid, Smartse, Skomorokh, Sonicsuns, Ncmvocalist, Aninhumer, January2007, Karjam, Tttom, Unused000702, Happysailor, JohnnyMrNinja, Martarius, C xong, Someone another, Arutoa, Addbot, MuZemike, Yobot, Yngvadottir, AnomieBOT, Earboxer, Mraellis, IManOM, Cnwilliams, GBev0213, LrdDimwit, Alpha9beta7, Josve05a, Erianna, Sirciny, BG19bot, MusikAnimal, BattyBot, Frosty, DarwinAnim8or, Lemnaminor, Shicky256, TWongNew, VictoryLap and Anonymous: 58

- **CherryOS** *Source:* https://en.wikipedia.org/wiki/CherryOS?oldid=607709661 *Contributors:* Ellmist, Plop, Whkoh, Fuzheado, HappyDog, Furrykef, Samsara, Mazin07, Psychonaut, Sverdrup, Rrjanbiah, AlistairMcMillan, OverlordQ, Ukexpat, Chmod007, M1ss1ontomars2k4, Pak21, KneeLess, MattTM, Evice, Minghong, CyberSkull, Pbones, Jchillerup, Euphrosyne, Cyro, Nuno Tavares, Woohookitty, Formeruser0910, Pastelink, Rjwilmsi, Brentm, Baojia, X1987x, Bash, FlaBot, Bovineone, Luckygod~enwiki, BiH, One, NetRoller 3D, SmackBot, Can't sleep, clown will eat me, Gh02t, Rogerbrent, Courcelles, RaviC, Zeke pbuh, Cydebot, Thijs!bot, John254, Pipedreamergrey, Puellanivis, R'n'B, Cyberfishee, 718 Bot, Proxy User, SF007, C. A. Russell, Addbot, Lightbot, Yobot, ClueBot NG, Raoouul, Gpatswiki and Anonymous: 44

- **Cipher Complex** *Source:* https://en.wikipedia.org/wiki/Cipher_Complex?oldid=666879333 *Contributors:* Rholton, Mandarax, Sdornan, Hibana, RussBot, N. Harmonik, OrphanBot, SkyWalker, Mika1h, Cydebot, Alaibot, BetacommandBot, Lord Hawk, Silver Sonic Shadow, Silver Edge, MarshBot, RobJ1981, Timkovski, CommonsDelinker, Delitas91, Darth Mike, Jrcla2, Varnent, Beem2, Monkeys rule 99, Jecht 91, SieBot, Tds247, TaerkastUA, Addbot, Dawynn, MuZemike, 威因, Enmoku, FrescoBot, DrilBot, FearfulforIP, Postwar, Enigma13278, ChrisGualtieri, Raymond1922A, DangerousJXD and Anonymous: 19

- **Demons of Mercy** *Source:* https://en.wikipedia.org/wiki/Demons_of_Mercy?oldid=677393324 *Contributors:* Gtrmp, Ron Ritzman, Koavf, Bgwhite, Pagrashtak, N. Harmonik, SmackBot, Tirkfl, RobJ1981, Beem2, Brycab, Someone another, Juliediet, Yobot, DrilBot, Skyerise, Mjs1991 and Anonymous: 5

- **Development hell** *Source:* https://en.wikipedia.org/wiki/Development_hell?oldid=686783750 *Contributors:* AxelBoldt, TwoOneTwo, The Anome, Grouse, Shsilver, SimonP, Comte0, Michael Hardy, Dante Alighieri, Justin Johnson, Mark Foskey, Rossami, JamesReyes, Dcoetzee, WhisperToMe, Furrykef, K1Bond007, Tempshill, Dale Arnett, Pigsonthewing, Merovingian, Der Eberswalder, Litefantastic, Andrew Levine, PBP, DocWatson42, Gtrmp, Sashal, Angmering, PeruvianIlama, Kabulykos, Varlaam, Xinoph, Gracefool, LucasVB, The Singing Badger, Jdl32579, Girolamo Savonarola, DragonflySixtyseven, Bodnotbod, TJSwoboda, Adambondy, Discospinster, Rich Farmbrough, Luvcraft, Barista, Ahkond, MajorB, Martpol, Bender235, Crooow, Kaszeta, CanisRufus, Surachit, PhilHibbs, Triona, Coolcaesar, TMC1982, Adambro, Bobo192, Deathawk, Cmdrjameson, Cwolfsheep, Steveklein, Hooperbloob, AndromedaRoach, Anthony Appleyard, CyberSkull, Fuzzybunny566, Ashley Pomeroy, InShaneee, Spangineer, Titanium Dragon, Kelson Vibber, Kesh, BanyanTree, Erik, Simon Dodd, Xayma, Zxcvbnm, Lkinkade, Feezo, Googleaseerch, Bacteria, Woohookitty, Mindmatrix, Havermayer, Uncle G, Headcase88, Commander Keane, Tabletop, Hailey C. Shannon, Hbdragon88, GregorB, Pfalstad, Keeves, JIP, Schmendrick, Deshem, Edison, Josh Parris, Rjwilmsi, Angusmclellan, Seraphimblade, Lilinka, SMC, Nick R, FuriousFreddy, Whitetigah, CR85747, Nidonocu, Elmer Clark, Mitsukai, Karrmann, SteveBaker, Hibana, Visor, Gdrbot, Igordebraga, Bgwhite, Bartleby, Satanael, RussBot, Peter S., Stephenb, Gaius Cornelius, Neilbeach, Zequist, Rhindle The Red, NawlinWiki, Irk, Smash, BigCow, KindOfBlue, Dumoren, SigPig, Martin villafuerte85, Musteval, ShadowMan1od, Mikeblas, Iicatsii, Empty2005, Adaru, Warrenm, CLW, Kelly42, AdamFunk, PTSE, Rfsmit, Nikkimaria, Alakazam, Th1rt3en, Toddgee, Livitup, Thermaland~enwiki, ArielGold, Sugar Bear, Whouk, PurplePlatypus, Jonathan.s.kt, Meegs, Bwfc, Kicking222, Mr. ATOZ, SmackBot, Joltman, Scott197827, Renegadeviking, Herostratus, Cavenba, Power piglet, Jagged 85, Verne Equinox, Cooksey87, Darklock, Exukvera, Shan246, Loompyloompy313, Master Deusoma, Canonblack, Portillo, The Famous Movie Director, Cabe6403, GoldDragon, Jnelson09, Thumperward, DoctorMud, Robocoder, RexImperium, Thatcrazycommie, Patriarch, TrackZero, Modest Genius, OneVeryBadMan, Tedronai, Britmax, New World Man, COMPFUNK2, Dyamantese, Napalm Llama, Cybercobra, Tiki2099, Quixada, Derek R Bullamore, Badgerpatrol, Ramrod?, Bob Castle, Wizardman, Kendrick7, TenPoundHammer, Vasiliy Faronov, Chaotic Mind, Axem Titanium, Delf, The jc, Drummer070, BrianInAtlanta, Rm w a vu, Karmus, Stanhead, Fritos~enwiki, Doczilla, TPIRFanSteve, Sonic Shadow, Wwagner, Violncello, Masem, Iridescent, Enter Movie, Poechalkdust, JoeBot, MGlosenger, Bebopblue, Woodshed, Eluchil404, Tawkerbot2, Billastro, Flubeca, Alexander Iwaschkin, CmdrObot, Gellarsgrudge, Killheart, Zarex, AlbertSM, Bsharkey, Mika1h, Rulesdoc, Melicans, Outriggr (2006-2009), Paulreilly86, Vectro, JDX, Otto4711, Lugnuts, Arrowned, Dematt, Irishwriter34, Ss112, SymlynX, Manfroze, DumbBOT, DiScOrD tHe LuNaTiC, Thenewestdoctorwho, AndTheCrowdGoesWild, Synetech, Arcayne, Lid, Barticus88, Deusfaux, W.A.C., PerfectStorm, Leedeth, Nalvage, Saruwine, Gmfs139, JustAGal, Marumae, RichardVeryard, Agent-Peppermint, PolarisSLBM, Silver Edge, Chegis, OptimumTaurus, Centrepull, MegaTroopX, General Norris, Jayron32, Mdotley, Dylan Lake, Eddyspeeder, Hoponpop69, Storkk, Ingolfson, Husond, Tony Myers, Wesborland, AvatarZero, Minnaert, MB1972, Albany NY, 100110100, IFCAR, Lighthope, Magioladitis, Vanished user ikjefknm34, MartinDK, Milesgray, Elcapitane, JaffaCakeLover, Interrobamf, Ryandsmith, Shorelander, Gwern, Asokoly, 989 RVD, Rettetast, Keith D, PF4Eva, Pacdude9, Markhh, Zlama, LevelNth, SCB '92, RoyBatty42, Andy5421, Grosscha, Anonywiki, GreenStapler, Renegade Replicant, PMBO, Floaterfluss, EyeOpener0013, Ras29, Potatoswatter, Pendo 4, Joost de Kleine, Sub Zenyth, Jackxmudd, Temppe, Speciate, Wikieditor06, Sonicsean89, Prede, TravellerDMT-07, Tyler9xp, Taraborn, Mercurywoodrose, The Big Eye, Walor, Jgiron, JayC, Someguy1221, Imasleepviking, UnitedStatesian, Random user 38749912, Bearian, Kazantakis60, Sarcha 45, TheValentineBros, Lamro, Vitz-RS, Universaladdress, Showers, DarthBotto, DrewSears, Joshshapiro, Winnd, DasBooch, GreaterWikiholic, RHodnett, NighthawkRider, TJRC, Claycrow, Mungo Kitsch, Yintan, Paulbrock, Ode2joy, Happysailor, Flyer22 Reborn, Peterlmccall, Oxymoron83, Lightmouse, Flukesh, TaerkastUA, Dabomb87, JL-Bot, YSSYguy, The Unknown Muncher, Martarius, Brain seltzer, Mezigue, William L Wright, Niceguyedc, Trivialist, DragonBot, Kanguole, Arjayay, Yonskii, Dthomsen8, Alexius08, Tustin2121, Addbot, American Eagle, MrZoolook, Jafeluv, Justallofthem, Download, SoloWing3844, Tassedethe, Nolelover, Jarble, Xowets, BlueMario1016, Ben Ben, Yobot, Granpuff, Donfbreed, Lacrymocéphale, GamerPro64, Angsc09, Mdw0, AnomieBOT, Hairhorn, Saginaw-hitchhiker, Materialscientist, Adam the silly, FrescoBot, VSfan88, Nanowolf, Holiverh, Secret Saturdays, Mundilfari, Christoph Balasse, RjwilmsiBot, John of Reading, AmericanLeMans, Tallungs, STATicVapor, Racerx11, Infobiac, GoingBatty, AStikhin, MikeyMouse10, Postwar, Josve05a, Unreal7, BNSF1995, Pote2639, EdoBot, Rock-O-Jello, ClueBot NG, Mansmokingacigar, Skjoldbro, DSnWiiRocks, JordoCo, Darthpineapple401, BG19bot, Nathan2055, StrangeApparition2011, Shirudo, CAWylie, BattyBot, CrosswalkX, ChrisGualtieri, ZappaOMati, Khazar2, Packer1028, Hannahvidal, Darth Molo, Ccamello1, Dexbot, Mogism, Smaxam, Damien Nguyen-Tran, Jfgoofy, GmanSir, Charmbox, Reverse polish, SomeFreakOnTheInternet, Jamesmcmahon0, Coltens14, Beastlcharizard13'sdefinitions, Toxophilus, Soffredo, LumCel, Gregnes2000, BenStein69, Lgnlint, Ilovetopaint, Intrepidusk, Game4brains, Justin.Parallax, Jakewhyland, Monkbot, TropicAces, MarkoPhoenix, TD712, MrWonka, ComicsAreJustAllRight, 3 of Diamonds, Rhpace, DanTaller, Sonic678, Tpy09352, AnimeDisneylover95 and Anonymous: 588

- **Duke Nukem 3D: Reloaded** *Source:* https://en.wikipedia.org/wiki/Duke_Nukem_3D%3A_Reloaded?oldid=675303024 *Contributors:* KAMiKAZOW, WikiPediaAid, Jtalledo, Rehevkor, WaltCip, Mika1h, Soetermans, Headbomb, Oosh, Geniac, Nightwheel, Ken g6, Varnent, TheValentineBros, Victory93, Triwbe, Martarius, Czarkoff, WulfenMortys, Saizo7, Darkcat1, Barballs, Eik Corell, XLinkBot, Addbot, NeD80, MuZemike, Yobot, AnomieBOT, Gap9551, Armbrust, PiistOff, Orppa, Ripchip Bot, Tomvanbraeckel, Thecheesykid, Jj98, ClueBot NG, Shaddim, Northamerica1000, Engineguyman, ChrisGualtieri, U'et and Anonymous: 51

- **Duke Nukem Forever** *Source:* https://en.wikipedia.org/wiki/Duke_Nukem_Forever?oldid=685969816 *Contributors:* Bryan Derksen, Roadrunner, Leidolf, Mrwojo, Frecklefoot, Chinju, Tregoweth, CesarB, Ahoerstemeier, KAMiKAZOW, CatherineMunro, Bueller 007, Salsa Shark, Kwekubo, Conti, Wooster, Clipdude, Dcoetzee, WhisperToMe, Ewald, Furrykef, Nv8200pa, K1Bond007, Quoth-22, AnonMoos, Rossumcapek, Nufy8, Robbot, Paranoid, Moncrief, Lowellian, Ashley Y, Wereon, BovineBeast, Xanzzibar, DraQue Star, Michael2, Alexwcovington, Centrx, DocWatson42, Jacoplane, DaB., Haeleth, Geeoharee, Ævar Arnfjörð Bjarmason, Avalean, KewlioMZX, Ssd, Primigenus, Bovlb, Naufana, BigHaz, RDevz, Iceberg3k, SoWhy, Piotrus, Testforechozero, Oneiros, Pmanderson, Sam Hocevar, Cynical, Ukexpat, Gerrit, Chmod007, Kasreyn, Gazpacho, SYSS Mouse, Tsman, Perey, Wfaulk, Erc, Discospinster, Rich Farmbrough, Qutezuce, Pie4all88, Martpol, Cyclopia, Scumbag, Fataltourist, Evice, PlasmaDragon, Project2501a, Kwamikagami, Claviola, Alereon, Mr. Strong Bad, Thunderbrand, Causa sui, Devil Master, Thisuser, Deathawk, Archfalhwyl, Sludge, VBGFscJUn3, Douglasr007, Joeldixon66, Cyrillic, Jason One, Matt Yohe, Bob rulz, Alansohn, Gary, Gargaj, Gamebrain, Ghostalker, Free Bear, Alyeska, CyberSkull, Ashley Pomeroy, Derumi, T-1000, Kristian Joensen, Angelic Wraith, Teggles, Wtmitchell, Ronark, TheRealFennShysa, DarkMythril, Tony Sidaway, Amorymeltzer, Drat, Bsadowski1, Pauli133, Zootm, Drbreznjev, Redvers, New Age Retro Hippie, Azertus, Cristan, Berserker79, ReelExterminator, Zntrip, Dopefish, Feezo, Jonloovox, CygnusPius, Woohookitty, Mindmatrix, TigerShark, LOL, MattGiuca, NeoChaosX, Tabletop, FoxInShoes, Hbdragon88, Al E., JFrosty, Bluemoose, GregorB, Sega381, Kralizec!, Pfalstad, CronoDAS, Dysepsion, Dweekly, Randomlypoetic, Xizer, Jclemens, Rjwilmsi, Seidenstud, Koavf, Fwend, Edholden, DynSkeet, Teklund, Sdornan, NeonMerlin, ElKevbo, Assassingod, Cakedamber, Brighterorange, AceTracer, Subtlesnake, Yamamoto Ichiro, Wragge, FlaBot, Master Thief Garrett, Wangoed, Fragglet, SuperDude115, Czar, DevastatorIIC, Jonny2x4, Huntersquid, Skierpage, BradBeattie, Snailwalker, Boyinabox, Sherool, Mhking, Igordebraga, Agamemnon2, Banaticus, Klingoncowboy4, ThunderPeel2001, Meridius, Koveras, I need a name, Hairy Dude, Retodon8, Torinir, Hede2000, Jellocube27, Hellbus, Cpc464, AstrixZero, Gaius Cornelius, XX55XX, NateDan, Pseudomonas, Gustavb, NawlinWiki, Mipadi, The Merciful, Complainer, Grafen, Snkcube, Clam0p, PatCheng, Pwlodi, Anetode, ShadowMan1od, LucentPhoenix, RUL3R, LodeRunner, Davidpk212, EEMIV, Falcon9x5, Captain Koloth, Vlad, Xino, N. Harmonik, Crisco 1492, Analoguedragon, Richardcavell, CharlieWiederhold, Charlie Wiederhold, Conglacio, Icydesign, PseudoKirby, Closedmouth, Th1rt3en, JuJube, ZoFreX, JoanneB, Shawnc, Piecraft, Syndrome~enwiki, Heavy bolter, ViperSnake151, Bluezy, Katieh5584, Captain Cornflake, Valters, Maxamegalon2000, Some guy, Jon Ace, Rehevkor, Aresmo, RichF, Ryūkotsusei, Riotgear, NetRolller 3D, Krótki, Gundam Bass, Kicking222, BPinard, SmackBot, Amcbride, Evilgrug, Bearda, InverseHypercube, McGeddon, David.Mestel, Sillygostly, Icarus Warrior, Lengis, Underwater, Nil Einne, Inonit, HalfShadow, David Fuchs, Master Deusoma, Peter Isotalo, Portillo, Ohnoitsjamie, Doktor Wilhelm, Oscarthecat, Maracle, Chris the speller, Happywaffle, Thumperward, Jgera5, Bignole, RexImperium, CSWarren, Lenin and McCarthy, Dethme0w, Can't sleep, clown will eat me, David Morón, Shalom Yechiel, OrphanBot, Zazpot, Xyzzyplugh, Thomas Connor, Sigmafactor, Xichael, Ferr, Sloverlord, Cybercobra, 1337 r0XX0r, Blake-, Chargh, Danjewell, Slawterr, PsychoJosh, AHOBOT, Superducktoes, Nickecb, Nairebis, Sonic Hog, Marcus Brute, Iniquity602, Kukini, Ged UK, TenPoundHammer, Ohconfucius, Blurion, CardinalFangZERO, Rklawton, Axem Titanium, Pizzahut2, TurboCat, Tktktk, Teancum, Mgiganteus1, ArgsIn2, Aleenf1, Stratadrake, A. Parrot, Stwalkerster, Mincetro, Bollinger, Yoduh1, Waggers, MarphyBlack, TPIRFanSteve, Kworn, Calysma, TJ Spyke, Masem, SubSeven, MikeWazowski, Unico master 15, Nehrams2020, Cat's Tuxedo, Iridescent, Peter M Dodge, Wjejskenewr, Sander Säde, Dp462090, Pata1, Steneub, Parkar, GarethEvans, FatalError, SkyWalker, Tifego, JForget, CmdrObot, Wikipedian06, 3d engineer, FunPika, Zarex, Dycedarg, Cyrus XIII, Goromike, Rockn-Roll, Mreleganza, Mika1h, NisseSthlm, NickW557, NTDOY Fanboy, Sir Lothar, Deusnoctum, Dusk83, Mesosade, Frankly Man, Shultz IV, A. Exeunt, Arrenlex, Cyberfray, Malamockq, Cydebot, Wikien2009, The Ultimate Koopa, Devletbek, Crossmr, Michaelas10, Gogo Dodo, DuckFerret, Corpx, Sxjxcx, Myscrnnm, Soetermans, DumbBOT, Superbowlbound, Bryan Seecrets, Inkington, Omicronpersei8, BetacommandBot, Thijs!bot, Epbr123, GeneralDuke, Hervegirod, Cipriano, Leftysrevenge, Blah3, Headbomb, Newton2, Mord3n, Marek69, Kathovo, X201, Tofof, The Hybrid, Mule Man, Krevans, SHivs, Silver Edge, Bbopman, Oreo Priest, The PC Gamer, CaptainGetts, AntiVandalBot, Howton92, Seaphoto, Pretender2j, MetaManFromTomorrow, Fnerchei, Afristelic, Blm07, Jj137, RobJ1981, BigSciZot, Z-vap, Fireice, Tws45, Daggoth, Dreaded Walrus, Markthemac, Kkmic, HanzoHattori, The Bloody Rose, Davewho2, Roman à clef, Dimension31, MER-C, Ribonucleic, Vlady24april, Tom Danson, Suoerh2, TheGoose109, J Greb, East718, Toglenn, LeJimster, Cnmapr, Y2kcrazyjoker4, Bencherlite, VoABot II, VerasGunn, Ambrosia-, AtticusX, Davidjk, Michael.m.winters, Jeron Moore, Aawood, ShadowTao, Prestonmcconkie, JaffaCakeLover, Pkaulf, Little Jimmy, Calvero2, Chris G, JaGa, Philg88, Link 486, Antondiep, Patstuart, Varsindarkin, Chimpanzee, Kiminatheguardian, Cube b3, Iamthereal, Alphamone, Kronnang Dunn, Grandia01, Superheat, Lord Einar, Killa Koz, Motley Crue Rocks, Armac, R'n'B, CommonsDelinker, Verdatum, Ccs4ever, Wbrice83186, Mausy5043, J.delanoy, Sirhenryjonesthe8th, Trusilver, Svetovid, Calamity-Ace, Richiekim, Silverxxx, Eliz81, Castef, Icseaturtles, Calvins48, Acalamari, Dispenser, Duhman0009, Gripdamage, Martyx, NinaOdell, TSCTH, Blackgaia02, Manofiorn, Sportz103, Tatrgel, White 720, Tiggerjay, SixteenBitJorge, Useight, Zephyr103, Varnent, Raidramon0, Halmstad, Steel1943, CardinalDan, Overcow, Gogobera, Vranak, Nikthestunned, Shortride, Jeff G., STLocutus, JamesTheWolf, Helenalex, Oshwah, Drhtl, Davehi1, The Sky May Be, Comrade Graham, Ryan shell, Mephista, SquirrelJS667, BrainSpecialist, Sam the Fish, Agent Q, Kurtpark, Clucko, Thelatemail, Verbatim9, Zchri9, DaSjieb, LeaveSleaves, Scargums, Master Bigode, EnigmaV8, Bllasae, VZakharov, Sharpie92, Michaeldsuarez, Geshpenst, Footballer121, Falcon8765, Enviroboy, Milkman519, Victory93, Adzma, Mantisia, Jimmi Hugh, Logan, Yerolo, Taylor975, Fjern, Dogsmellerz1, Harley Quinn hyenaholic, Ryguy88games, BobHackett, Froo, BotMultichill, PanagosTheOther, Krawi, Sothicus, One more night, Yuefairchild, Swaq, Crash Underride, Keilana, Bentogoa, LeGUIGUI, Nebulousity, MattParker 119, I destroy wiki, Chemtype, Spock2266, AnonGuy, Lightmouse, Casablanca2000in, Keiron waites, Miqrogroove, Jericho1337, DavidDW, Tds247, Dwinches, Gunmetal Angel, Gyrferret, Onopearls, Backin5minutes, Info845, ShackGrudgeMatch, Dravecky, Spitfire19, Silvergoat, Legionarius, Bamadio, Peniche5, Cragscleft, TaerkastUA, Darkranger-red, Dabomb87, Ekerazha, Kashakak, Wonchop, FeldsparQartz, Goodsblame1, Zanderdude55, ImageRemovalBot, Wikipedian-Marlith, Martarius, ClueBot, Txpete, Strongsauce, GrandDrake, PipepBot, Elephant Talk, Fyyer, The Thing That Should Not Be, Starkiller88, Hong620, Krogstadt, Skäpperöd, JTBX, Pineapplehead14, LizardJr8, Cirt, LukeTheSpook, Gordon Ecker, TMV943, Canis Lupus, Jusdafax, Pablo jre, Revelatipsum, Troy A. Stark, John Nevard, Muhandes, Austen791, Millionsandbillions, MJDTed, Krazymike, Kauzio, Thinggar, Thingg, Aitias, Nicolaibo~enwiki, Diogo Pinto, Snoozer282, Party, Phantomwiki, DumZiBoT, Chiefchips, Dominic Neagle, XLinkBot, Tuxlie, Will-B, Ost316, WikHead, BlackDeath3, NellieBly, Noskap, Fiskbil, RedXnoimage, Thereaper14, Addbot, Slykiller7, JBsupreme, Some jerk on the Internet, The Other Saluton, DougsTech, RG4ever, Mr. Wheely Guy, ContiAWB, Darkness2005, Fluffernutter, Damiens.rf, Ismouton, Dyadron, Skyezx, TristramPW, Ld100, Nickin, SoloWing3844, Favonian, Herr Gruber, Rich1631, Tide rolls, Lightbot, Byakuya Truelight, Krano, NeD80, Gail, ScienceApe, Xowets, Howcouldi, Nongame, MissAlyx, Htews, Ben Ben, Luckas-bot, Yobot, Drychicken, Nissarana, Redranger241, Asskiser123, Angsc09, Plasticbot, Linfocito B, AnomieBOT, KiasuKiasiMan, ShadowRatchet92, DarkLight748, MLVD, Galoubet, Berntie, Piano non troppo, Khroz2rdrez, Radardude2007890, Materialscientist, Citation bot, Christaylor80, Bread climper, Frankenpuppy, TUFKAAP, ThomasSixten, Dead Echo DX, Mistbite, Saintsrow2, 4twenty42o, Christianb5, Gensanders, Ched, TheFireTones, Frosted14, OccasionalAcidFlashBack, Anhydrobiosis, Plmr, Wikieditor1988, Smallman12q, FaTony, Basileias, Liamandjerm, VasOling, Bam050196, Dougofborg, Deltasim, Griffinofwales, Razr95, ESJBond007, Sunstriker, Babyboy2689, TheOccasionalAcidFlashbake, FrescoBot, Jajohnsen, Szili32, Ringerfan23, Ndboy, Minooch, JMS Old Al, ParaDoxus, GrislyGrizzly, HJ Mitchell, DelphinidaeZeta, CakePwns!, Jackfail, S2 Lovely Boy, Victor.spain, Purpleturple, Xhaoz, FonFon Alseif, HamburgerRadio, Dkandyoshi, Swedhombre, Dgre005, Liorti92, Hellknowz, Calmer Waters, Darbacour, RedBot, Dwigs, WhoaThereTiger, Ptarjan, Full-date unlinking bot, Mlbiggs101, Something12356789101, Typhoon966, CuddlySatan, StevenMario, Doxy66, WesUGAdawg, Veldrik, JuhoV, Reaper Eternal, Diannaa, Tbhotch, Rabbitshadow, Noodlesgc, Gmantonz, Brakoholic, RjwilmsiBot, Bento00, ButOnMethItIs, DarkApollo, Phlegat, Shadow16nh, Poohze, DASHBot, CR4ZE, John of Reading, WikitanvirBot, Obamafan70, Avenue X at Cicero, Gfoley4, Heymid, Honza007, GoingBatty, Dotarray, Billy the Worm, Lollymann, Wikipeli, Sanitized Handitizer, Biotech45, Thecheesykid, Postwar, ZéroBot, John Cline, Fæ, Strahan201, Neurochild, Shanesimms, Ό οἶστρος, Alshaheen15, Bilbo571, Codeisle, Stuckmanx9, MAINEiac4434, Dukenukemforverftw, Funnysheep, Trikeen, MissingNoLLL, Sbmeirow, Veritas4ever, Τασουλα, Bugandhoney, IGeMiNix, FrankFlanagan, L Kensington, Smackerlacker, Damirgraffiti, ShrimpMonkey, ChuispastonBot, Wikimakesmart, DukeNukemForever2020, TheDeviantPro, Scratchy96, Mikitei, MigrantP, ShatteredSpiral, Haloreachking666, Smidgens, Hollowkubox, Faramir1138, Freemanukem, ClueBot NG, Mike088, Aaron Booth, Jack Greenmaven, Horseman16, Korbiin, DoctorFernungli, EcsTactic, Mewashere1, Robthepiper, Broden, TruPepitoM, DukeNukemFan, Conman2011, Ramasti, Flyingnarb, Rahulghose, Cntras, VillaYouth,

HonzaG, Easy4me, Megadave94, Phantomjerad, Adwiii, Aallasdfa67usgd60, Letsips, Thegamemuster, Shovan Luessi, Ambition44, Crazymonkey1123, Oddbodz, Rndomuser, Helpful Pixie Bot, Flannely, Tommy758, 254Jackson, Rangeshifter, Cheerleadersammy, Pangaeo, Casualgaymer, Fazm1bico, J to the Dog, Ckywht, Negativecharge, Nicholiserviaphd, Kyledehaven, George.kinsman, The Last Username I Could Think Of, CityOfSilver, PTJoshua, Supermidget98, DukeNukemNuk, FAKEACCOUNT1234, DukeNukemNuk2, Mckenna1994, ElitezX, JoeFission, Jasper1000, Artamentix, DarkGreen revenge, HowardCoward, Trolledsoftly, BlooPigeon, Lutari88, Frze, Yoshidome, Compfreak7, Tinktinkbubbles, Jamezrr, Marveilleux, C-Fresh790, Asdwdhgdg, Lolzilla98, Datstuph, Daggerfallfreak12, Assmongler, Bevo4pres1, Engineguyman, Airpr23, MrJeffreyOrlando, Pratyya Ghosh, ChrisGualtieri, Jwilliam85, YFdyh-bot, Fuckoffsir, SNAAAAKE!!, SteelSkin667, BDE1982, Frosty, Andyhowlett, KahnJohn27, ILOVEFRASIER, EvergreenFir, TheWikiDuke, Sarahfleck, Albert11317, NottNott, Ginsuloft, Fred8229, Murricaman, Anoymus lol, Daß Wölf, Machetemaster32, AdrianGamer, DangerousJXD, 302ET, Gamingforfun365, TrialMacameau, SPQR93 and Anonymous: 1663

- **Fluorescent multilayer card** *Source:* https://en.wikipedia.org/wiki/Fluorescent_multilayer_card?oldid=537972261 *Contributors:* Camembert, Nazo~enwiki, Radiojon, Mike Storm, Rich Farmbrough, .:Ajvol:., Ringbang, Linuxbeak, Robertvan1, Cholmes75, Teply, SmackBot, Dancter, TheBendster, JL-Bot, Lightbot, Erik9bot, TheLou75 and Anonymous: 6

- **Fluorescent Multilayer Disc** *Source:* https://en.wikipedia.org/wiki/Fluorescent_Multilayer_Disc?oldid=562095053 *Contributors:* Malcolm Farmer, Heron, Camembert, Nazo~enwiki, Rl, Denni, DocWatson42, Ryanrs, Edcolins, Neilc, Toytoy, Ehudshapira, Urvabara, Evice, .:Ajvol:., Fritzpoll, BRW, Paul1337, Grenavitar, Gene Nygaard, Ringbang, Zntrip, Fbriere, Vegaswikian, Fish and karate, FlaBot, Crazycomputers, Sherool, Retodon8, Bamgooly, The1physicist, Hugh Bennett, Abune, Eptin, SmackBot, Tsca.bot, Sonic3KMaster, CmdrObot, Schmloof, PrestonH, CzarNick, 2help, Synthebot, TheBendster, ImageRemovalBot, Snigbrook, SpikeToronto, XTerminator2000, DumZiBoT, Addbot, LaaknorBot, Lightbot, Legobot II, Cureden, Smartie2thaMaxXx, EdoBot, Gary Dee and Anonymous: 34

- **Glaze3D** *Source:* https://en.wikipedia.org/wiki/Glaze3D?oldid=575417554 *Contributors:* Chowbok, Pauli133, Daveydweeb, Kbdank71, Ketiltrout, MZMcBride, RussBot, Ataqer, SmackBot, Agentbla, Hmains, Mdwh, Frap, Jesse Viviano, Gioto, Reedy Bot, Rilak, DanielPharos, ChrisGualtieri and Anonymous: 9

- **Godus** *Source:* https://en.wikipedia.org/wiki/Godus?oldid=688151585 *Contributors:* Pgrote, Auric, John Hubbard, Master Thief Garrett, Rwalker, CapPixel, Mika1h, Cydebot, Danrok, Soetermans, Odie5533, Magioladitis, Nuuskamikkonen, Frmorrison, Yobot, AnomieBOT, IrrtNie, Lacon432, ClueBot NG, BG19bot, ConCelFan, Aisteco, The1337gamer, BattyBot, Madmax108, Dja1979, ChrisGualtieri, Leitoxx, Noxbird, ArmbrustBot, Monkbot, Diraffe, NotNowImMining and Anonymous: 46

- **The Grinder** *Source:* https://en.wikipedia.org/wiki/The_Grinder?oldid=677853451 *Contributors:* Misterkillboy, Closeapple, Mahanga, Koavf, Bgwhite, 2fort5r, Super Mario, Geoff B, Waggers, LostOverThere, Cydebot, X201, Unknownsage13, Little Jimmy, Varnent, Bovineboy2008, Beem2, Henryodell, Exert, Epass, TaerkastUA, WikipedianMarlith, Martarius, XLinkBot, Wyatt915, Addbot, The Editor 155, Tide rolls, Yobot, AnomieBOT, Armbrust, GrislyGrizzly, DrilBot, Lightlowemon, Enaz909, CyberTiger531, Awsome8, H3llBot, ClueBot NG, Awesomeness95, Reg porter, Bmf 051 and Anonymous: 71

- **Haystack (software)** *Source:* https://en.wikipedia.org/wiki/Haystack_(software)?oldid=671366252 *Contributors:* Open Source Guy, Siroxo, Wk muriithi, Diego Moya, MiG, Koavf, Clerambj~enwiki, Vegaswikian, Strolls, SmackBot, Frap, Npdoty, Gigi head, ARC Gritt, Tchickdp, Mercurywoodrose, Rorysolomon, Addbot, Dawynn, Damiens.rf, G8briel, Yobot, Shootbamboo, FrescoBot, LucienBOT, Trankvila, Mono, Smartiger, Kamran the Great, Karthikndr, ResidentAnthropologist, Lucas Boullosa, Aturpat2015 and Anonymous: 8

- **Hellraid** *Source:* https://en.wikipedia.org/wiki/Hellraid?oldid=687527943 *Contributors:* Soetermans, X201, Varnent, Beem2, Martarius, Niceguyedc, Khudhayyer, XLinkBot, Addbot, Dawynn, The Editor 155, AnomieBOT, CoolingGibbon, GoingBatty, Lacon432, Fazm1bico, SNAAAAKE!!, Dissident93, Cryptic games, TCMemoire, Landingdude13, Zombiefan99, AdrianGamer, Zombiefan1999, DangerousJXD, Srednuas Lenoroc and Anonymous: 23

- **Highlander: The Game** *Source:* https://en.wikipedia.org/wiki/Highlander%3A_The_Game?oldid=685683513 *Contributors:* Liftarn, Bgwhite, Rehevkor, SmackBot, Master Deusoma, Bignole, CardinalFangZERO, SkyWalker, Mika1h, Cydebot, TOOTCB, UberMan5000, X201, Silver Edge, Pixelface, Xeno, Magioladitis, Aengusnokitsune, A Nobody, Rosenknospe, Jmodum90, Beem2, Andrew22k, Stormin' Foreman, TaerkastUA, Mr. Granger, Martarius, Inuboy1000, Mild Bill Hiccup, MrKIA11, Uncle JonBob, HooperBandP, Shikakimush, Cgohanm, Lightbot, AnomieBOT, Shotgunemmet, Rushwz, Joaquin008, ASOTMKX, DrilBot, Full-date unlinking bot, Ashutomar1994, GoingBatty, H3llBot, Treylander, Petrb, Antiqueight, Assasaint, Dissident93, Paleface Jack and Anonymous: 50

- **Holographic Versatile Disc** *Source:* https://en.wikipedia.org/wiki/Holographic_Versatile_Disc?oldid=672835368 *Contributors:* Rjstott, Jtoomim, Leandrod, Modster, Gabbe, Ahoerstemeier, KAMiKAZOW, Haakon, Ciphergoth, Conti, Rob.derosa, Tempshill, Thue, BenRG, Northgrove, Phil Boswell, Psychonaut, Kwi, Blainster, Jondel, Netjeff, Mattflaschen, David Gerard, Giftlite, Peruvianllama, Zaphod Beeblebrox, Gracefool, Bobblewik, Pythagoras1~enwiki, 159753, Toytoy, Bosmon, Tooki, Hellisp, Ukexpat, Uzisuicide, Canterbury Tail, Danh, Blorg, Urvabara, Rich Farmbrough, Rhobite, FT2, Sn0wflake, Kbh3rd, Evice, Pjrich, PhilHibbs, RoyBoy, Foobaz, Giraffedata, Renwique, Alansohn, CyberSkull, Atlant, Moocowred, Paul1337, Danhash, Mikeo, Gene Nygaard, Ringbang, NuVanDibe, Dismas, Smokeala, Feezo, Thryduulf, Ixistant, Brazil4Linux, Andreas -horn- Hornig, Mb1000, NazismIsntCool, Jon Harald Søby, Cbyneorne, Synkronos, Sin-man, David Levy, Koavf, Gabrielsimon, Arabani, KamasamaK, FlaBot, RexNL, Preslethe, Chobot, Raider Duck, Mhking, Rotsor, Wjfox2005, YurikBot, RobotE, Icedemon, Shawn81, Kb1koi, Emersoni, PrimeCupEevee, Wangi, Marktaff, Empty2005, Closedmouth, Abune, CapitalLetterBeginning, Peter, PureLegend, SigmaEpsilon, Garion96, NetRolller 3D, Sardanaphalus, Locke Cole, SmackBot, T2k, TheWickerMan, Jrockley, Thunder Wolf, Nil Einne, Iamthebob, Keegan, Lordkazan, Oli Filth, Hibernian, Hmich176, Nintendude, Sashafcb, Can't sleep, clown will eat me, OrphanBot, TheKMan, Alyssa3467, NoIdeaNick, Beetle120, BlueCube, Dreadstar, WillV, Demicx, Phontain, Soumyasch, MattEngland~enwiki, Leigh.bardsley, NJZombie, NongBot~enwiki, Lucid, Swotboy2000, Muhaidib, JoeBot, CP\M, Tawkerbot2, Shortgeek, Sketch051, Randhirreddy, Ravensfan5252, Nick Wilson, Gogo Dodo, Dancter, Anthonynow12, Benspar, BlueRaja, A3RO, Electron9, Alphius, AntiVandalBot, Luna Santin, Clarenceville Trojan, Prolog, Pseudo account, Outsid3r, JAnDbot, Inverse.chi, JogyB, PhilKnight, DataMatrix, Magioladitis, Mclean007, Edgelord, Turkishbob, AMK1211, Nikevich, Catgut, MetsBot, Chris G, Ztobor, Conquerist, J.delanoy, Uncle Dick, Sd31415, Sweetness46, Anonymous Dissident, Cody-7, Bagle, Bertrem, Grendelum, Pjoef, Thunderbird2, Ageton, SieBot, Mrxg4, Foljiny, Editore99, Momo san, Lightmouse, Priyatanuroy, Jimmy Slade, Martarius, ClueBot, Mpdimitroff, Eric Wester, Gopher65, DragonBot, Alexbot, Jusdafax, Thehelpfulone, La Pianista, Возрождение, Dr. Omar Al-Oraby, Nblschool, Kasey 670, Glacier Wolf, Malevolution, Antonyh3, Addbot, Ghettoblaster, Xx521xx, SpellingBot, Soapfoam, West.andrew.g, Lightbot, Luckas-bot, Glatisant, O Fenian, Captainrag, JDC808, AnomieBOT, Piano non troppo, ArthurBot, Xqbot, Kagemaru16, Capricorn42, Drilnoth, Derolic, Kyng, NoJr0xx, LucienBOT, Kavankvsit, MathFacts, Meaning of Lif, Cufflinks 010, Flint117, Smuckola, Full-date unlinking bot, DanielL5583, Blue Em, Renegadeviking2, Diannaa, Thex1, Born2bgratis, Joshua.Li, EmausBot, Danny856, Kokken Tor, Caspertheghost, Smartie2thaMaxXx, Youhd, ClueBot NG, SenseiAC, BG19bot, Togfan, HiDefRev, Patrick87, Bumblebritches57, ArmbrustBot, Comp.arch, Granhil and Anonymous: 397

- **Huxley (video game)** *Source:* https://en.wikipedia.org/wiki/Huxley_(video_game)?oldid=652859215 *Contributors:* K1Bond007, Nufy8, Dj245, Pak21, Thunderbrand, Bobo192, Disastrophe, Cmdrjameson, Pearle, Bob rulz, CyberSkull, Snowolf, Stephan Leeds, TigerShark, Mathmo, Al E., Kralizec!, Alastair~enwiki, Mrbartjens, BD2412, Rjwilmsi, Sdornan, Harro5, JP Godfrey, Ems57fcva, Sion Delta, Hibana, Hahnchen, PatCheng, Thiseye, Brandon, Bobquest3, N. Harmonik, Slicing, MLA, Tik~enwiki, Shawnc, ArielGold, Jedi6, TLSuda, Draconus, Hide&Reason, Kicking222, SmackBot, Reedy, Deiaemeth, Bluebot, A. B., KieferSkunk, Addshore, WhereAmI, Sajman12, John, Kiyobi, Estarrol, Ehheh, Spiffyxd, Cmsr. Jackdaw, Obtsu, Unico master 15, Major Master, K0r3aN pR1d3, JStewart, Courcelles, SkyWalker, Shmargin, CmdrObot, Mika1h, Randalllin, Esko~enwiki, BF2MCguy, (chubbstar), Cydebot, Crossmr, Sora45, Tawkerbot4, DumbBOT, Omicronpersei8, BetacommandBot, Epbr123, Lord Hawk, X201, Nick Number, Fogeltje, AntiVandalBot, David136a, Zedla, DJBaku, MER-C, Andonic, Tartaros, Planb11, VoABot II, MastCell, Brusegadi, Make, Starayo, Pax:Vobiscum, Neo Geo, Jackson Peebles, Rettetast, J.delanoy, Eliz81, AntiSpamBot, Wincrest, ThatGreenMachine, Sd31415, Hakkahakkabazoom, Sa5r, Infinity inc, Signalhead, Yellosnolvr, Station1, Brodvold, Nicolegetoffmyleg, Beem2, Red117, Mrix2000, Mueslitm, John.n-irl, Gobbly2100, XA-9, Moozaad, Keilana, UnrivaledShogun, Cheesewedge, COBot, Bean23, BURNyA, Rcc coolio, Wolfcub11, Desensitize, ImageRemovalBot, ClueBot, Dnguyen1319, Element926, TimRaver, Contralya~enwiki, Eric Wester, Toeface401, Hauzer-Ninja, Bellum et Pax, Ecobun, Nymf, Commdor, Thingg, DumZiBoT, The Phantomnaut, Eik Corell, XLinkBot, Reizah, Ost316, Abandoned, MystBot, Wyatt915, Danxm, Addbot, Philipsucks26, Megata Sanshiro, Jimmy Figsworth, D0762, Tide rolls, Lightbot, MuZemike, AnomieBOT, Deathwiki, Blueflamez, Piano non troppo, Ikkyone, Ulric1313, E2eamon, Guitarbarian, LilHelpa, 1wolfblake, Motorheadx, Jmundo, Saad92, PigFlu Oink, SL93, Seahorseruler, Minimac, Phlegat, Salvio giuliano, John of Reading, MigrantP, Handelabra, Doubledose 2, SNAAAAKE!!, Ranze, BlitzGreg, Hakken and Anonymous: 239

- **Hyper CD-ROM** *Source:* https://en.wikipedia.org/wiki/Hyper_CD-ROM?oldid=576554023 *Contributors:* Jun-Dai, Woohookitty, Jeff3000, FlaBot, SmackBot, Ohconfucius, Lightmouse, Tbsdy lives, Addbot, Sfaefaol, FrescoBot, Helpful Pixie Bot and Anonymous: 10

- **Imperator Online** *Source:* https://en.wikipedia.org/wiki/Imperator_Online?oldid=648354902 *Contributors:* Steinsky, Stewartadcock, Alan De Smet, Saucepan, Kuralyov, Sam Hocevar, Thunderbrand, HolsteinCow, Bob rulz, Gary, CyberSkull, Messup, Marasmusine, LizardWizard, Zenoseiya, ADeveria, JIP, RobyWayne, XX55XX, Aaron Brenneman, N. Harmonik, Nerrolken, Robofish, Amalas, JohnCD, Imperator3733, JohnnyMrNinja, Aelfnig, GoingBatty, BG19bot and Anonymous: 15

- **Jet Thunder** *Source:* https://en.wikipedia.org/wiki/Jet_Thunder?oldid=664666964 *Contributors:* Carnildo, Marasmusine, BD2412, LjL, FlaBot, SmackBot, Colonies Chris, Phantom brave, Frap, X201, Davehi1, ImageRemovalBot, Polly, MatthewVanitas, Addbot, Kman543210, Yobot, AnomieBOT, DrilBot, NinjaTazzyDevil, Faramir1138, BattyBot, Mogism and Anonymous: 10

- **Killing Day** *Source:* https://en.wikipedia.org/wiki/Killing_Day?oldid=666144402 *Contributors:* SmackBot, X201, ACSE, Beem2, Technopat, ScienceApe, Yobot, IShadowed, Whywhenwhohow, AvicAWB, Nath gamer, BG19bot and Anonymous: 8

- **Kingdom Under Fire II** *Source:* https://en.wikipedia.org/wiki/Kingdom_Under_Fire_II?oldid=688447679 *Contributors:* YUL89YYZ, RoyalFool, Gary, Woohookitty, Mandarax, RussBot, Falcon9x5, N. Harmonik, Josh3580, Edders, SmackBot, MK8, SkyWalker, Mika1h, Jpaluf, Cydebot, Crossmr, Soetermans, X201, RobJ1981, Pixelface, Barek, Xeno, Geniac, Keith D, Anonymous Keeper of Records, Castef, Beem2, RSEagle, Michaeldsuarez, Allmightyduck, StaticGull, ImageRemovalBot, ClueBot, WarAnakin, Auntof6, Dcwil477, Dakovski, Versus22, XLinkBot, Nrbnerd17, Ost316, Addbot, Poco a poco, Megata Sanshiro, Zeta Nova, AnomieBOT, 威因, LilHelpa, CoolingGibbon, 1wolfblake, Levis93, PS3 Delita, Dead at hell but alive, Unknown198283, Leeh7, DrilBot, Full-date unlinking bot, Lightlowemon, Reaper Eternal, RjwilmsiBot, Smappy, Afatatlot, Lacon432, Bilbo571, Orangemohawk, MigrantP, ClueBot NG, Imgaril, Glacialfox, Klslsk, Neofirezero, Mddkpp, BattyBot, Jgrassilli, Ajsitian, Oranjblud, Pcgamer17, Firorianxia, Dissident93, Sealkhk, Martin464646, Sayiamina, Bl1zz4rd-editor, Prof.Haddock, Oxcide, Cue The Corruption, Monkbot, Landingdude13, AdrianGamer, Zacharyalejandro and Anonymous: 94

- **Kirby's Return to Dream Land** *Source:* https://en.wikipedia.org/wiki/Kirby'{}s_Return_to_Dream_Land?oldid=684327842 *Contributors:* Kapow, M1ss1ontomars2k4, Wild Bill, Neko-chan, Apostrophe, CyberSkull, Diego Moya, Tsuba~enwiki, New Age Retro Hippie, Red dwarf, Firsfron, SwordKirby537, Rjwilmsi, Koavf, Zooba, Gakon5, Hibana, Khisanth, BlinksTale, Xoloz, Kimchi.sg, Smash, Pagrashtak, N. Harmonik, Karuma, TITROTU, Bly1993, Doom127, Rehevkor, SmackBot, ShadowRanger, KnowledgeOfSelf, Indium, Parrothead1983, Jnelson09, George Ho, Can't sleep, clown will eat me, Unknown Dragon, ShadowUltra, LtPowers, Drumpler, Hope(N Forever), IronGargoyle, 041744, Retromaniac, MisterCQNZR, Grandy02, TPIRFanSteve, TJ Spyke, Cipher, Judgesurreal777, TheHorseCollector, RoboCafaz, Mika1h, Stevo1000, Rockysmile11, Cydebot, Cambrant, Sora45, DumbBOT, Alaibot, Xubelox, Widkid85, BetacommandBot, Kirbman101, Jjam189, 0dd1, Marek69, Kirbykook, SNS, Coconutfred73, X201, Seaphoto, RobJ1981, Jhsounds, Vendettax, Kung Fu Man, Kariteh, ThomasO1989, Chaoshi, Justin The Claw, CTF83!, JeffreyChau, Sonicxtreme, Disaster Kirby, Chimpanzee, CommonsDelinker, Zaiise, DanJ, Sapphire Flame, Bowser Jr. Koopa, Funguy100, L'Aquatique, Blackgaia02, Proto Dude, STBotD, Mario Characters, Ashura777, Casper10, Bowserjr55, Marioman12, Volcanux, Varnent, Johnny Au, Bovineboy2008, TXiKiBoT, X3ni, JayC, Zagnut, Kurowoofwoof111, Enviroboy, Dylanlip, Unused000702, Miremare, Vanished user 82345ijgeke4tg, GonbeFAN, Judicatus, GEM036, Mr.Detective, DragonZero, Blake, Yair rand, Wonchop, JesseMeza, ImageRemovalBot, Martarius, SupaPopTa!!!, Snrabu, Bubbletruble, Theseven7, Pink D3stroyer, WhereIsTheCite?, GENERALZERO, Mr.Mario 192, Super Shy Guy Bros., NuclearWarfare, BOTarate, Swindbot, Pikachu247, DumZiBoT, XLinkBot, Ost316, PhoenixMourning, Meatlkirby, Megapen, MystBot, Pupi18, Addbot, Megata Sanshiro, Debresser, Kaoruu, LemmeyBOT, Lightbot, MuZemike, Megaman en m, Ptbotgourou, AnomieBOT, Project-79, Jackie Stuntmaster, 威因, Materialscientist, Invincibob, Snorlax Monster, HalberdStopCrashing, DSisyphBot, GrouchoBot, Sergecross73, Macbookair3140, Shadowjams, Samwb123, Tttttttt999, Pigyman, Kirbywiimaster, Frest123, GrislyGrizzly, DrilBot, RedBot, SonicGamer, Ian10234, TheStrayCat, EmperorFishFinger, Reach Out to the Truth, RjwilmsiBot, Jay2518, Vellidragon, EmausBot, Gamer3059, WikitanvirBot, Notshane, Ajraddatz, 0030520dv, Daedae917, Naruto1419, Stickboy52697, Tehaura, Wikipelli, Thecheesykid, ZéroBot, King kong 9000, Thomasdav, Railer-man, LatinoSeuropa, Jennet Uraida, LikeLakers2, ClueBot NG, MelbourneStar, Satellizer, Zeroslashj, Chester Markel, Sonofthecastle, Easy4me, ScottSteiner, Mtking, Newimagekirby, Gabriel Yuji, Vanished user ivweij23ijwefk4, MusikAnimal, Vaati the Wind Demon, Nogginfan, SuperELLG, TheBaehr, Kirbykid20, PhilipTerryGraham, Dentalplanlisa, 22dragon22burn, Khazar2, Interlude65, Mastergamer1091, Brony :3, Bacon-Cheddar Man 5000, Cobalion254, SamX, Richolmes14, Monkbot, Papyrus-winged ninja Akil, Landingdude13, Pokedu, AdrianGamer, ClassicOnAStick, Allen7054, Zacharyalejandro, Therealsonix41 and Anonymous: 338

- **Mercs Inc** *Source:* https://en.wikipedia.org/wiki/Mercs_Inc?oldid=667566678 *Contributors:* Mika1h, Cydebot, Bailmoney27, Popa01, A Nobody, Beem2, Jack Merridew, Flyer22 Reborn, Nymf, WölffReik, Some jerk on the Internet, Soundout, Nohomers48, Nickin, AnomieBOT, KiasuKiasiMan, Materialscientist, F-86, Tbhotch, Crunchman2600, Skamecrazy123, Acather96, Postwar, Bilbo571, The Country Girl, Shakinglord, Michaelmas1957, Dissident93, Jamesbondfanforever, Landingdude13 and Anonymous: 22

- **Mya (program)** *Source:* https://en.wikipedia.org/wiki/Mya_(program)?oldid=688508186 *Contributors:* Rich Farmbrough, Nikkimaria, Brianyoumans, Mr Stephen, Spinningspark, P. S. Burton, Legobot, Yngvadottir, Freikorp, Starship.paint, Mogism and Belle

- **Noctis** *Source:* https://en.wikipedia.org/wiki/Noctis?oldid=673877706 *Contributors:* EALacey, Ramir, Luigi30, Sonjaaa, Airconditioning, DMG413, Lightning4, Illarkul, GregoryWeir, Saxifrage, Marasmusine, Woohookitty, Bluemoose, Euchrid, Rjwilmsi, Teemu Maki~enwiki, Ian Pitchford, Hibana, Slicing, Theodolite, Plankhead, Babij, SmackBot, McGeddon, Faya, Klichka, Colonies Chris, Synapse001, Gohst, Ten-PoundHammer, Ravimakkar, Wickethewok, Eridani, Gregbe, Mika1h, Mzk, MetalGearLiquid, Jac16888, Cydebot, Pascal.Tesson, BetacommandBot, Keb25, Jonathan Williams, Xeno, Granpire Viking Man, Blood Oath Bot, Alessandro Ghignola, Wingedsubmariner, Nave.notnilc, Mazzbeast, James.Denholm, Mingamango181, MrKIA11, Alexbot, Conical Johnson, Randomran, AzraelUK, Eik Corell, Addbot, Grayfell, Yobot, AnomieBOT, Drilnoth, Molokaicreeper, DrilBot, Jaguar, Full-date unlinking bot, Lorson, Jfmantis, Harbingerdawn, ClueBot NG, Nodulation, Samwalton9, Several Pending, 93, GabeIglesia, Strath104, Jefequeso, HyperspaceCloud and Anonymous: 67

- **Octopiler** *Source:* https://en.wikipedia.org/wiki/Octopiler?oldid=544304381 *Contributors:* Chowbok, Yurik, Malcolma, Samuel Blanning, Addbot, TheAMmollusc, Cwa210, ZéroBot and Anonymous: 4

- **The Outsider (video game)** *Source:* https://en.wikipedia.org/wiki/The_Outsider_(video_game)?oldid=679064495 *Contributors:* Ary29, MBisanz, Thunderbrand, CyberSkull, Drat, Stemonitis, GregorB, Holek, Sdornan, Bgwhite, RussBot, Bill, SmackBot, Ominae, Oscarthecat, Nareek, Master Rex, SkyWalker, Mika1h, Wykebjs, Tawkerbot4, Alaibot, BetacommandBot, X201, Silver Edge, Chrisjj3, Cyclonius, Oetzi101, Magioladitis, Aawood, R'n'B, Delitas91, GrahamHardy, Doceirias, Davehi1, SieBot, 1836311903, TaerkastUA, ClueBot, EoGuy, Addbot, Dawynn, Colt9033, Lightbot, Yobot, Captain Cheeks, Thatguy1010, DrilBot, Illegitimate Barrister, Helpful Pixie Bot, 1morey and Anonymous: 26

- **Prey 2** *Source:* https://en.wikipedia.org/wiki/Prey_2?oldid=679111301 *Contributors:* Bearcat, Junkyardprince, DNewhall, Mahanga, Zntrip, Sdornan, Master Thief Garrett, RussBot, NateDan, Cryptic, Matticus78, Rehevkor, SmackBot, Chargh, Ohconfucius, SkyWalker, Mika1h, Sir Lothar, Malamockq, Soetermans, Deathmonkey7, Leftysrevenge, Gamer007, SNS, X201, Dbrodbeck, Geniac, Magioladitis, JPG-GR, Equazcion, Izno, Nikthestunned, VolkovBot, Beem2, Comrade Graham, Oxfordwang, EonOmega, Yintan, RabisaE, JohnnyMrNinja, TaerkastUA, Martarius, ShawnIsHere, Nymf, Frank Rawland, Moozipan Cheese, Addbot, Tassedethe, Timothygormley, NeD80, NeoBatfreak, Yobot, DarkLight748, Jim1138, Xqbot, CoolingGibbon, Berni2k, CR4ZE, Jaklink, Postwar, XXAntibodyXx, Bilbo571, SporkBot, Jman98, The-HeronGuard, MigrantP, Freemanukem, ClueBot NG, Wugu, Control-202, Jiggylypiggly, Helpful Pixie Bot, Fazm1bico, AwamerT, Timur9008, ConCelFan, RscprinterBot, Egret Of Regret, BattyBot, SideMaster, Sivos909, Æ−202, Dissident93, Yash!, SloppyVagFlaps, Blabla5678, Epicgenius, Secondhand Work, Lifeischanging, Prodfather, Iseesky, JimeoWan, EimaiTrelooos, Landingdude13, AdrianGamer, Chad Washington and Anonymous: 116

- **Project Xanadu** *Source:* https://en.wikipedia.org/wiki/Project_Xanadu?oldid=683560318 *Contributors:* Timo Honkasalo, The Anome, Stephen Gilbert, Nealmcb, XeoX, Kku, Cyde, Yann, Tregoweth, DavidWBrooks, Artost, Jackseay~enwiki, CatherineMunro, Jimregan, RodC, David Latapie, Greenrd, Lfwlfw, Bevo, Nyh, Wereon, HaeB, Adaptr, Brouhaha, Ewg, Elektron, Cynical, RalfZosel~enwiki, Vcool, Gazpacho, Pmsyyz, Byrial, Pavel Vozenilek, Shlomif, RoyBoy, Rpresser, Viriditas, Cwolfsheep, Foobaz, Syzygy, Physicistjedi, Ashley Pomeroy, Uucp, Tony Sidaway, LFaraone, Pol098, Tabletop, GregorB, Angusmclellan, Grlloyd, FlaBot, Margosbot~enwiki, YurikBot, Manop, Deodar~enwiki, Jpbowen, DuncanCragg, Nlu, TransUtopian, Extraordinary, Vicarious, MrBucket, SmackBot, John.ohno, Chronodm, Scott Paeth, Bluebot, Liebeskind, Meco, Simon12, George100, MikeWren, Cydebot, SymlynX, JamesBrownJr, Ratneshdeepak, Barek, Never been to spain, VoABot II, Dropframe, Gwern, Gjd001, Yaron K., Johnpacklambert, VirtualDelight, Arivelz, Olivianewtonjohn, Dispenser, DadaNeem, Hervold, Squids and Chips, The Wild Falcon, Bsroiaadn, AlleborgoBot, Kraquehaus, Svick, Peter Walt A., Alexbot, Arjayay, C. A. Russell, Addbot, Mabdul, Jarble, Legobot, Luckas-bot, Yobot, Ecobiotics, AnomieBOT, Yblumberg, Tekks, RibotBOT, MastiBot, John of Reading, Splibubay, Bcaulf, Maximilianklein, Mikhail Ryazanov, Snotbot, Helpful Pixie Bot, Autodidaktos, Khanklatt, Tleaver, QuietJoon, Cube00, Monkbot, Vieque, Chocolatechip65, Konlyn and Anonymous: 51

- **Protein-coated disc** *Source:* https://en.wikipedia.org/wiki/Protein-coated_disc?oldid=675190396 *Contributors:* Northgrove, Gracefool, OldakQuill, 1297, Tooki, FT2, Evice, Joe 042293, ArielGold, Sardanaphalus, King Nintendoid, Bluebot, SuperBuuBuu, George Church, Can't sleep, clown will eat me, CoolKoon, Romeu, Cydebot, Anthonynow12, Will Bradshaw, IndigoAK200, Prolog, Magioladitis, Ishikawa Minoru, Edgelord, Acertain, Gwern, Manticore, Varanwal, Remmus4, PixelBot, Thingg, DumZiBoT, Addbot, Lightbot, Luckas-bot, Baradys, Spectatorbot13, DrilBot, SporkBot, Smartie2thaMaxXx, BattyBot and Anonymous: 41

- **Rainbow Storage** *Source:* https://en.wikipedia.org/wiki/Rainbow_Storage?oldid=635789158 *Contributors:* Pnm, MacGyverMagic, Demiurge, Ouro, Rcog, Discospinster, Wtshymanski, Woohookitty, Dar-Ape, Docbug, Wongm, Wavelength, Ksyrie, Trek1701~enwiki, Cffrost, Luk, SmackBot, PaulWay, Neo-Jay, Nixeagle, LeContexte, Sameerb, Amitpnt, Tubezone, CharlotteWebb, Barek, Time3000, Public Menace, Princess Tiswas, Mstuomel, Mrzeevi, Sainul, Jeff G., Maghnus, ClueBot, XLinkBot, Addbot, Lightbot, Smallbrady89, Sz-iwbot, Ashok.tcr, Full-date unlinking bot, Sapphirewhirlwind, Rotten regard and Anonymous: 39

- **Sadness (video game)** *Source:* https://en.wikipedia.org/wiki/Sadness_(video_game)?oldid=682284125 *Contributors:* Timwi, Altenmann, OverlordQ, Deathawk, CyberSkull, ReyBrujo, Culix, New Age Retro Hippie, Deansfa, Karam.Anthony.K, Marudubshinki, Leden, Koavf, FlaBot, Zooba, UnlimitedAccess, Sceptre, Oni Lukos, N. Harmonik, Calaschysm, TITROTU, SmackBot, State of Love and Trust, Prototime, Hmains, Chris the speller, Enbob89, Unknown Dragon, Bendragonbrown47, SevereTireDamage, Nakon, Raezr, Lukemicallef, Tehw1k1, Ck lostsword, Salty!, Hope(N Forever), -al, EddieVanZant, Liam Plested, Masem, Sabrewing, Audiosmurf, Mika1h, Rastusa, Leemsy, Cydebot, Thebigguylp, LordHuffNPuff, Gogo Dodo, Dancter, Alaibot, Quetzalcoatl45, Ebyabe, Deathmonkey7, BetacommandBot, Danhm, Lord Hawk, X201, Opelio, Scepia, RobJ1981, JackSparrow Ninja, ThomasO1989, 100110100, Dekimasu, Rich257, PrestonH, DanJ, Castef, SharkD, AntiSpamBot, Belovedfreak, M0dded, STBotD, Flamesplash, Slydevil, Gatotsu911, Biyyt, Bice, Maxim, Zeiros, C45207, Astoc, Weeman com, Miremare, Nuttycoconut, Sceptile2390, ImageRemovalBot, Martarius, ClueBot, Emperor Ing, Wiimaster07, Bbtsikito, Lbrun12415, Aileza, Kakofonous, Eik Corell, BarretB, MystBot, Addbot, Next-Genn-Gamer, The Editor 155, Megata Sanshiro, Lost on Belmont, Troopapunk, Mr T (Based), AarnKrry, GrislyGrizzly, DrilBot, Johnnykamikaze, Reach Out to the Truth, Michael553, ClueBot NG, Satellizer, Poseidon1224, TheLoverofLove, SNAAAAKE!!, Arron rowsky and Anonymous: 136

CCHIPSS, Dabomb87, Cowman2333, KSven, MenoBot, ClueBot, Strongsauce, Frmorrison, Piledhigheranddeeper, Another Believer, N3me$i$, Bashum104, Saeed.Veradi, Ost316, GenosBrawl, BobyMcBobbob, Gazimoff, Addbot, Tassedethe, Tide rolls, Lightbot, Yobot, GamerPro64, KiasuKiasiMan, Citation bot, LilHelpa, The Banner, Tdrotim, Juwy12345, Leon3289, Dinamik-bot, Alex IW12, Baskirs777, Electroguv, H3llBot, ClueBot NG, The Master, Lowercase sigmabot, BG19bot, Wooloomooloo1, Wiki13, Mdann52, SNAAAAKE!!, Br'er Rabbit, Ranze, Bobfrazer, Confusing123456789, Serpinium, Landingdude13, DangerousJXD, Tasuki37, Ghost Van Dal and Anonymous: 255

- **Stonekeep** *Source:* https://en.wikipedia.org/wiki/Stonekeep?oldid=679748860 *Contributors:* Frecklefoot, Edward, DavidWBrooks, Rursus, Iosif~enwiki, Brian Kendig, Grutness, Alai, WilliamKF, Firsfron, Jeff3000, ADeveria, Pictureuploader, Alrik Fassbauer, Phantom784, Bgwhite, Bwaquin, N. Harmonik, Analoguedragon, SmackBot, EwokCommanda, Edgar181, Grumpyyoungman01, TJ Spyke, SubSeven, Judgesurreal777, MagikGimp, CmdrObot, Mika1h, Senorelroboto, Cydebot, Acolyte of Discord, Odie5533, BetacommandBot, Thijs!bot, Voracious reader, Mojo Hand, Nick Number, Milton Stanley, Nimdok, HanzoHattori, Magioladitis, Threedots dead, R'n'B, Gatotsu911, TXiKiBoT, Comrade Graham, Billinghurst, UnneededAplomb, OberRanks, Bagatelle, Lisatwo, Rosiestep, Peteroid, JohnnyMrNinja, KingCapybara, ClueBot, PipepBot, Sfisher2, Panicpunk, XLinkBot, Anticipation of a New Lover's Arrival, The, Addbot, Captain Obvious and his crime-fighting dog, Victoriaearle, AnomieBOT, Grey ghost, MASLEGOMan, J04n, Deltasim, FrescoBot, KaterBegemot, Schmeater, Salvidrim!, Full-date unlinking bot, Martin IIIa, Agraff91687, John of Reading, OnePt618, Tyros1972, Wbm1058, Mike Agricola, SNAAAAKE!!, Boskee, Mogism, Monkbot, Macrowriter and Anonymous: 47

- **Team Fortress 2** *Source:* https://en.wikipedia.org/wiki/Team_Fortress_2?oldid=686182420 *Contributors:* Scipius, Voidvector, Jogloran, Furrykef, Nv8200pa, Jeoth, Jeffq, Nufy8, Bkell, 75th Trombone, Reid, Nifboy, Icenine0, HelgeHan, Orangemike, VPeric, Michael Devore, Golbez, Alanl, Neilc, Nova77, Plutor, Sonjaaa, Chris Ducat, Aeonite, Ukexpat, Aknorals, M1ss1ontomars2k4, Trevor MacInnis, Qjuad, EagleOne, Discospinster, Avriette, Qutezuce, Tristan Schmelcher, Ahkond, Bender235, ESkog, Kyz, Mr. Strong Bad, RoyBoy, Dennis Brown, LordRM, WhiteTimberwolf, Thunderbrand, Causa sui, Bobo192, Pikawil, CoreyEdwards, Conny, Alansohn, Gary, Gargaj, CyberSkull, SlimVirgin, Echuck215, Bart133, Radical Mallard, TaintedMustard, Fourthords, ReyBrujo, EAi, Lapinmies, Drat, BlastOButter42, Axeman89, New Age Retro Hippie, Bobrayner, Woohookitty, NeoChaosX, WadeSimMiser, GregorB, Kralizec!, AndrewNeo, Prashanthns, Spread, A3r0, Holek, Mandarax, RadioActive~enwiki, Vanderdecken, Rjwilmsi, Koavf, Sdornan, Voretus, Miserlou, Brighterorange, Toby Douglass, Reinis, Yar Kramer, Fish and karate, FlaBot, Xero, SchuminWeb, RexNL, Czar, Kri, BradBeattie, Liontamer, Chobot, Bgwhite, Hahnchen, Sceptre, I need a name, Taurrandir, Mustkillroy, Gregalodon, RussBot, IanManka, MrCheshire, Powerlord, Gaius Cornelius, NateDan, Rsrikanth05, TonicBH, Daveswagon, NawlinWiki, Dysmorodrepanis~enwiki, Theadept, Snkcube, Tom Edwards, Topperfalkon, FivePointPalmExplodingHeart, The Obfuscator, Brandon, Ravedave, Boadrummer, Moe Epsilon, Froth, Saberwyn, R.D.H. (Ghost In The Machine), Charron, Rosensteel, EEMIV, Falcon9x5, BOT-Superzerocool, FiggyBee, PyroGamer, N. Harmonik, Pizzahut~enwiki, J. Nguyen, Zzuuzz, Mahalis, Bongomanrae, Gtdp, Homepie, Closedmouth, Clayhalliwell, Th1rt3en, Fergofrog, CWenger, Momus, Smurfy, Skeith, Zquack, Rehevkor, Lomacar, CIreland, Bibliomaniac15, Luk, Sikyanakotik, Phinnaeus, SmackBot, EvilCouch, Djkor, KnowledgeOfSelf, McGeddon, Sillygostly, Pielover87, Dxco, Jab843, ZS, ComputerSherpa, Mintpieman, David Fuchs, Master Deusoma, Cheshil, Oscarthecat, The monkeyhate, Stuart P. Bentley, Chris the speller, Bluebot, BarkerJr, Jftaylor21, Jprg1966, Corinthian, TheFeds, Busterdawg, Emperor Jachra, Kostmo, Takua108, Darth Panda, Blueshirts, Mrbutter, Rama's Arrow, Wingspantt, Pedroshin, Mr.hotkeys, Daniel-Dane, Rrburke, AttackingHobo, DurotarLord, Addshore, Celarnor, Treygdor, Nayl, Nahum Reduta, Minty Fresh Death, Esin0420, Logan GBA, Earl CG, Bonecrushah, Watcher95, Dreadstar, Warren, DMacks, Einhanderkiller, S@bre, AbsoluteFlatness, BobbyPeru, Pilotguy, Ged UK, Ohconfucius, Pakundo, Blastoboy1000, Krashlandon, IgWannA, Harryboyles, Valfontis, Pizzahut2, ZenSaohu, E3l, Chumba1, J 1982, BurnDownBabylon, Marksda, Katstevens, Robofish, BlisteringFreakachu, Arnoox, Joffeloff, Minglex, Cyberlink420, JamesWeb, IronGargoyle, Farazparsa, Stratadrake, Taotd, Plunge, Beetstra, SQGibbon, AxG, Snl99, MarphyBlack, Maddox, Jdng, Ryulong, Legonovsky, Lunajurai, Masem, ShaleZero, Cat's Tuxedo, Hypergeek, Iridescent, HertzaHaeon, Skorp, Mario 64 Master, Blakegripling ph, Vocaro, Twas Now, JustinRossi, Courcelles, WakiMiko, Crazydog115, Jacono, SkyWalker, Daedalus969, ShakespeareFan00, JForget, CmdrObot, Ivan Pozdeev, Zarex, Addict 2006, IsaCanuck22, Mika1h, Freaky Dug, Minglong, GhostStalker, Herman238, El aprendelenguas, FlyingToaster, Sir Lothar, LectR, Frankly Man, Andkore, H.M.S Me, Dogman15, Grenno, Cydebot, Gogo Dodo, Dsbos, Schmatz~enwiki, Badpazzword, Myscrnnm, Soetermans, Shirulashem, Aanhorn, Chuto, DumbBOT, Quadrius, Happy Lollipop, Guyinblack25, Ward3001, Mfko, Tooga, Hopex, EdenMaster, LilDice, Ayzmo, BetacommandBot, Thijs!bot, JAF1970, Epbr123, Izer~enwiki, N5iln, BenXO, Gamer007, Mojo Hand, Headbomb, Marek69, Evan1109, James086, X201, Marcotulio, Sciss0rz, MasterDeva, Thebenefactor, Nick Number, Barracuda1187, AbsenceWiki, Dawnseeker2000, CTZMSC3, Salavat, KitAlexHarrison, AntiVandalBot, Uselesswarrior, Fnerchei, Dvandersluis, DannyB!!, MECU, Spencer, Oddity-, Pixelface, Fearless Son, Nimdok, Dreaded Walrus, Ingolfson, Kariteh, Knowsitallnot, Oskard, GaMeRuInEr, Planetary, Vangoghs Ear, Frostbitex460, Xeno, Qwo, Bigserg, ColoradoCNC, DanPMK, ProjectPlatinum, Stuffed Lizard, Y2kcrazyjoker4, Acroterion, DataMatrix, Magioladitis, Schwans, Dp76764, VoABot II, Squiddy7, Da Mainman, RMN, Blakeo x, JamesBWatson, Jéské Couriano, Avicennasis, JaffaCakeLover, Janizdreg, Fabrictramp, Solid Runner, ChaosPrime, Zaybertamer, Animum, Pkaulf, Gordonfreeman 19, Torchiest, XMog, Wonton, DerHexer, Esanchez7587, WLU, Hypo Mix, Stephenchou0722, Neo Geo, Katana314, ChibiVegito, Cmsjustin, Smashman2004, R'n'B, Kahonee, Wikipope, Galahaut, Blindraven, Tgeairn, RockMFR, Shniken, J.delanoy, FANSTARbot, Tdreher, Bogey97, UBeR, Uncle Dick, Ginsengbomb, Eliz81, Extransit, Foober, A Nobody, Plankton5005, Smartalic34, Kenori, Icseaturtles, SharkD, Barts1a, Dispenser, BrokenSphere, Katalaveno, Geo24dude, McSly, Bailo26, Sliker Hawk, IROK, AntiSpamBot, Stratman, Iambob4444, CAF51, Raw bean, Marcellus2070, Phirazo, Amigaoasis, Ironfatdogs, Gutsnipe, Atropos235, RedDragon111, Cometstyles, FuegoFish, Vanished user 39948282, Evil Egg, Varnent, Vinsfan368, Gracz54, Rfdparker, Resplendent, Vranak, Vor'Cha, X!, Nitrowolf, Bullmeister, VolkovBot, Riahc3, Pleasantville, Jeff G., Cbigorgne, Andrewjomega, Black raven2525, Gunnar Guðvarðarson, EchoBravo, LeilaniLad, Philip Trueman, Nophysicalbody, TXiKiBoT, Fwuffy, Master Jaken, Hgtonight, Infinitymmx3, AJanuary, Iliketofrolic666, Arnon Chaffin, Zimbardo Cookie Experiment, Clucko, Modul8r, Orc2888, Jackfork, Themcman1, Sentineneve, CoupleKlonopin, SpecMode, DexterHolland, Madhero88, TheValentineBros, The Great Unwashed, Lova Falk, Falcon8765, Enviroboy, Hitari0, Suzaku Medli, Tiger97882, Sir Mad Hatter, Victory93, Unused0030, Nagy, Lyinginbedmon, 400Hz100V, Harley Quinn hyenaholic, SieBot, Lalaland111111, Chaosof99, Tresiden, Scarian, Weeliljimmy, VVVBot, Therightclique, Olivier Beaton, Phe-bot, Sophora, Yuefairchild, Yintan, Vanished user 82345ijgeke4tg, Krusty klamen, Sexykins13, Down M., Tiptoety, Hageba, The juggresurection, MyrddinE, Ferret, CaelumArisen, DevOhm, Oxymoron83, AngelOfSadness, Tombomp, Lyle Piratehog IV, CollisionCourse, Professional Waffle, Fillabunny, Rorsta, BURNyA, Jscorp, Wonchop, Kanonkas, Dammej, Necravolvr, WikipedianMarlith, ApokalypseCow, Mcge0053, Twelvepack, NNR07, Feeder18, Martarius, Sfan00 IMG, ClueBot, Totaltf2, LAX, Kamulako, GamerErman2001, AngelGraves13, Caboose218, Itskamilo, The Thing That Should Not Be, Mriya, Matdrodes, Vilepickle, Roberteliteno1, Thapban, EoGuy, TomRed, Lawrence Cohen, R000t, Theghoul22, Vergil 577, Nayr86, HiQu, Mads Ren`ai, Coolhandlueck, Washboardplayer, JTBX, KuroFalcon, Mezigue, WDM27, CounterVandalismBot, NovaDog, Niceguyedc, Iriath Zhul, Mydrian, MrBosnia,

Universe?oldid=683699848 *Contributors:* Maximus Rex, Eric42, Neilc, AstroBlue, Thunderbrand, Jason One, Zippanova, Y0u, Firsfron, Woohookitty, Tabletop, Gaius Cornelius, N. Harmonik, Red Jay, Ryūkotsusei, SmackBot, Dwanyewest, Master Deusoma, Tehw1k1, TJ Spyke, Mika1h, Neelix, Cydebot, X201, Mendinso, Jhsounds, Hut 8.5, Frogacuda, Master Bigode, Winchelsea, Ray and jub, Ost316, Megata Sanshiro, Lightbot, Anna Frodesiak, Nicktoonspl15, FrescoBot, Salvidrim!, Canyq, Souvalou, ClueBot NG, TServo2050, Bmusician, Slrfjgz, ShirleyM-cLoon, Jowee the Raposa, Greefan443 and Anonymous: 29

- **Ugo Volt** *Source:* https://en.wikipedia.org/wiki/Ugo_Volt?oldid=617013529 *Contributors:* Pagrashtak, SmackBot, Mika1h, JaGa, Diogo Pinto, SF007, Addbot, Dawynn, JoshuaD1991, Yobot, 威因, DrilBot, Orenburg1, Western John, Fazm1bico and Anonymous: 8

- **Warcraft Adventures: Lord of the Clans** *Source:* https://en.wikipedia.org/wiki/Warcraft_Adventures%3A_Lord_of_the_Clans?oldid= 680785090 *Contributors:* Shsilver, Wwwwolf, Sannse, Slug, Conti, Phil Boswell, Xanzzibar, Gtrmp, Reub2000, Ausir, Darkhunger, SoWhy, Lacrimosus, Alsocal, Yaz0r, CALR, Discospinster, Moochocoogle, Andrew Maiman, Bob rulz, CyberSkull, Bart133, Drat, New Age Retro Hippie, Falcorian, Mindmatrix, Nuggetboy, Thorpe, MattGiuca, Commander Keane, FoxInShoes, JIP, Imperialles, Canderson7, Staecker, FlaBot, CR85747, SpectrumDT, UnlimitedAccess, Sus scrofa, RussBot, Havok, PatCheng, Syrthiss, N. Harmonik, Avraham, DocEGM, Hide&Reason, SmackBot, Master Deusoma, Winterheart, Taric25, Poulsen, Chris the speller, DavidRolfe, Thumperward, Greatstoneplanet, Mordea, Ceoil, Yayaba, Ace of Sevens, Judgesurreal777, Monkfish1, SkyWalker, Mika1h, Hemlock Martinis, Soetermans, Medievaldragon, BetacommandBot, Supparluca, Headbomb, X201, Fogeltje, AntiVandalBot, DocVM, Canadian-Bacon, JAnDbot, MER-C, RicardoFachada, Dream Focus, Endomorphic, Atheuz, Varnent, Izno, TXiKiBoT, Synthebot, Horselover145, Rhickey1986, ImageRemovalBot, AlptaBot, CohesionBot, ImageBacklogBot, DumZiBoT, MystBot, Addbot, Megata Sanshiro, LaaknorBot, Lightbot, Limideen, Obersachsebot, CoolingGibbon, J4lambert, GrouchoBot, Dinamik-bot, Canyq, Andreona and Anonymous: 68

3.2 Images

- **File:360-91-panel.jpg** *Source:* https://upload.wikimedia.org/wikipedia/commons/9/9e/360-91-panel.jpg *License:* Public domain *Contributors:* ? *Original artist:* ?

- **File:Ambox_current_red.svg** *Source:* https://upload.wikimedia.org/wikipedia/commons/9/98/Ambox_current_red.svg *License:* CC0 *Contributors:* self-made, inspired by Gnome globe current event.svg, using Information icon3.svg and Earth clip art.svg *Original artist:* Vipersnake151, penubag, Tkgd2007 (clock)

- **File:Ambox_globe_content.svg** *Source:* https://upload.wikimedia.org/wikipedia/commons/b/bd/Ambox_globe_content.svg *License:* Public domain *Contributors:* Own work, using File:Information icon3.svg and File:Earth clip art.svg *Original artist:* penubag

- **File:Ambox_important.svg** *Source:* https://upload.wikimedia.org/wikipedia/commons/b/b4/Ambox_important.svg *License:* Public domain *Contributors:* Own work, based off of Image:Ambox scales.svg *Original artist:* Dsmurat (talk · contribs)

- **File:Amiga_Walker.png** *Source:* https://upload.wikimedia.org/wikipedia/en/c/c5/Amiga_Walker.png *License:* Fair use *Contributors:* http://www.blachford.info/computer/walker/walker.html *Original artist:* ?

- **File:Amiga_Walker_Motherboard.png** *Source:* https://upload.wikimedia.org/wikipedia/en/d/d1/Amiga_Walker_Motherboard.png *License:* Fair use *Contributors:* http://obligement.free.fr/articles/walker.php *Original artist:* ?

- **File:Aviacionavion.png** *Source:* https://upload.wikimedia.org/wikipedia/commons/6/68/Aviacionavion.png *License:* Public domain *Contributors:*

- Turkmenistan.airlines.frontview.arp.jpg *Original artist:* Turkmenistan.airlines.frontview.arp.jpg: elfuser

- **File:BGE2-scr1.jpg** *Source:* https://upload.wikimedia.org/wikipedia/en/2/22/BGE2-scr1.jpg *License:* ? *Contributors:* ? *Original artist:* ?

- **File:BGE2-scr2.jpg** *Source:* https://upload.wikimedia.org/wikipedia/en/8/85/BGE2-scr2.jpg *License:* ? *Contributors:* ? *Original artist:* ?

- **File:Black_Mesa-Surface_Tension.jpg** *Source:* https://upload.wikimedia.org/wikipedia/en/2/26/Black_Mesa-Surface_Tension.jpg *License:* Fair use *Contributors:* [1] *Original artist:* ?

- **File:Bob'{}s_Game_logo.png** *Source:* https://upload.wikimedia.org/wikipedia/commons/8/81/Bob%27s_Game_logo.png *License:* Public domain *Contributors:* http://images1.wikia.nocookie.net/bobsgame/images *Original artist:* Uploaded by MuZemike at en.wikipedia

- **File:Chaos_emeralds.svg** *Source:* https://upload.wikimedia.org/wikipedia/commons/f/f0/Chaos_emeralds.svg *License:* CC BY-SA 3.0 *Contributors:* Own work *Original artist:* Gringer

- **File:Cherry_os_logo.png** *Source:* https://upload.wikimedia.org/wikipedia/en/b/b9/Cherry_os_logo.png *License:* Fair use *Contributors:* The logo may be obtained from CherryOS. *Original artist:* ?

- **File:Commodore-64-Computer.png** *Source:* https://upload.wikimedia.org/wikipedia/commons/3/34/Commodore-64-Computer.png *License:* Public domain *Contributors:* Own work *Original artist:* Evan-Amos

- **File:Commons-logo.svg** *Source:* https://upload.wikimedia.org/wikipedia/en/4/4a/Commons-logo.svg *License:* ? *Contributors:* ? *Original artist:* ?

- **File:Crystal_Clear_action_run.png** *Source:* https://upload.wikimedia.org/wikipedia/commons/5/5d/Crystal_Clear_action_run.png *License:* LGPL *Contributors:* All Crystal Clear icons were posted by the author as LGPL on kde-look; *Original artist:* Everaldo Coelho and YellowIcon;

- **File:Crystal_Clear_app_kedit.svg** *Source:* https://upload.wikimedia.org/wikipedia/commons/e/e8/Crystal_Clear_app_kedit.svg *License:* LGPL *Contributors:* Sabine MINICONI *Original artist:* Sabine MINICONI

- **File:DukeNukemForever.jpg** *Source:* https://upload.wikimedia.org/wikipedia/en/0/0e/DukeNukemForever.jpg *License:* Fair use *Contributors:*
  The cover can be obtained from Gearbox Software and 2K Games. *Original artist:* ?

- **File:Dukesig.png** *Source:* https://upload.wikimedia.org/wikipedia/en/d/d6/Dukesig.png *License:* Fair use *Contributors:*
  http://www.dukenukemreloaded.com/forum/viewtopic.php?f=9&t=116&sid=5eb4a7712cb468959ac67c07d2f5c435 *Original artist:* ?

- **File:Edit-clear.svg** *Source:* https://upload.wikimedia.org/wikipedia/en/f/f2/Edit-clear.svg *License:* Public domain *Contributors:* The *Tango!*
  Desktop Project. Original artist:
  The people from the Tango! project. And according to the meta-data in the file, specifically: "Andreas Nilsson, and Jakub Steiner (although
  minimally)."

- **File:Emoji_u1f4be.svg** *Source:* https://upload.wikimedia.org/wikipedia/commons/f/fb/Emoji_u1f4be.svg *License:* Apache License 2.0 *Contributors:* https://code.google.com/p/noto/ *Original artist:* Google

- **File:Esther_Dyson_Monaco_Media_Forum.jpg** *Source:* https://upload.wikimedia.org/wikipedia/commons/2/2b/Esther_Dyson_Monaco_
  Media_Forum.jpg *License:* CC BY-SA 2.0 *Contributors:* http://www.flickr.com/photos/eirikso/3032265750/ *Original artist:* Eirik Solheim
  (flickr user: eirikso)

- **File:Folder_Hexagonal_Icon.svg** *Source:* https://upload.wikimedia.org/wikipedia/en/4/48/Folder_Hexagonal_Icon.svg *License:* Cc-by-sa-3.0
  Contributors: ? *Original artist:* ?

- **File:Gamepad.svg** *Source:* https://upload.wikimedia.org/wikipedia/en/b/be/Gamepad.svg *License:* ? *Contributors:* ? *Original artist:* ?

- **File:Girls_of_E3_2011_No.15.jpg** *Source:* https://upload.wikimedia.org/wikipedia/commons/6/69/Girls_of_E3_2011_No.15.jpg *License:*
  CC BY 2.0 *Contributors:* https://secure.flickr.com/photos/22904000@N07/5814271258/ *Original artist:* Richard Cabrera

- **File:Glaze3dwest1.jpg** *Source:* https://upload.wikimedia.org/wikipedia/en/9/9a/Glaze3dwest1.jpg *License:* ? *Contributors:* ? *Original artist:* ?

- **File:Gnome-searchtool.svg** *Source:* https://upload.wikimedia.org/wikipedia/commons/1/1e/Gnome-searchtool.svg *License:* LGPL *Contributors:* http://ftp.gnome.org/pub/GNOME/sources/gnome-themes-extras/0.9/gnome-themes-extras-0.9.0.tar.gz *Original artist:* David Vignoni

- **File:HVD_Logo_shaded.png** *Source:* https://upload.wikimedia.org/wikipedia/commons/b/b1/HVD_Logo_shaded.png *License:* Public domain *Contributors:* thic.org/pdf/Jul05/optware.mdeese.050719.pdf *Original artist:* HVD Forum

- **File:HVDstruct.svg** *Source:* https://upload.wikimedia.org/wikipedia/commons/2/2d/HVDstruct.svg *License:* GFDL *Contributors:* <a href='//commons.wikimedia.org/wiki/File:HVDstruct.png' class='image'><img alt='HVDstruct.png' src='https:
  //upload.wikimedia.org/wikipedia/commons/thumb/b/bd/HVDstruct.png/100px-HVDstruct.png' width='100' height='64' src-set='https://upload.wikimedia.org/wikipedia/commons/thumb/b/bd/HVDstruct.png/150px-HVDstruct.png 1.5x, https://upload.wikimedia.
  org/wikipedia/commons/thumb/b/bd/HVDstruct.png/200px-HVDstruct.png 2x' data-file-width='307' data-file-height='196' /></a> *Original
  artist:* Fred the Oyster

- **File:Half_Life-Surface_Tension.jpg** *Source:* https://upload.wikimedia.org/wikipedia/en/e/ec/Half_Life-Surface_Tension.jpg *License:* Fair
  use *Contributors:*
  Half-Life version 1.1.1.0 for Microsoft Windows.
  Original artist: ?

- **File:Highlandergame.jpg** *Source:* https://upload.wikimedia.org/wikipedia/en/d/da/Highlandergame.jpg *License:* Fair use *Contributors:*
  Taken from this website: http://www.xbox360fanboy.com/photos/highlander-1/596205/ *Original artist:* ?

- **File:Highlandergame2.jpg** *Source:* https://upload.wikimedia.org/wikipedia/en/c/ce/Highlandergame2.jpg *License:* ? *Contributors:* ? *Original
  artist:* ?

- **File:Hvd_disc.jpg** *Source:* https://upload.wikimedia.org/wikipedia/en/3/38/Hvd_disc.jpg *License:* Fair use *Contributors:*

- http://www.howstuffworks.com/hvd.htm *Original artist:* ?

- **File:Joystick_black_red_petri_01.svg** *Source:* https://upload.wikimedia.org/wikipedia/commons/d/d8/Joystick_black_red_petri_01.svg *License:* CC0 *Contributors:* ? *Original artist:* ?

- **File:Kirby_rtdl_wii.jpg** *Source:* https://upload.wikimedia.org/wikipedia/en/a/a2/Kirby_rtdl_wii.jpg *License:* Fair use *Contributors:*
  http://www.eurogamer.net/gallery.php?game_id=4581&article_id=1368147#anchor *Original artist:* ?

- **File:Logo_of_Six_Days_in_Fallujah.png** *Source:* https://upload.wikimedia.org/wikipedia/commons/8/8e/Logo_of_Six_Days_in_Fallujah.
  png *License:* Public domain *Contributors:* http://www.beautifulpeoplesclub.org/ *Original artist:* Atomic Games

- **File:Logo_of_huxley.png** *Source:* https://upload.wikimedia.org/wikipedia/en/6/61/Logo_of_huxley.png *License:* Fair use *Contributors:*
  The logo may be obtained from Webzen Games Inc.. *Original artist:* ?

- **File:Meet_The_Scout.jpg** *Source:* https://upload.wikimedia.org/wikipedia/en/b/bd/Meet_The_Scout.jpg *License:* ? *Contributors:* ? *Original
  artist:* ?

- **File:Mercs_Inc_Cover.png** *Source:* https://upload.wikimedia.org/wikipedia/en/7/77/Mercs_Inc_Cover.png *License:* Fair use *Contributors:*
  Electronic Arts
  Original artist: ?

- **File:Mya_program_character.png** *Source:* https://upload.wikimedia.org/wikipedia/en/2/24/Mya_program_character.png *License:* Fair use
  Contributors:
  Bizony, Piers (2001). Digital Domain: The Leading Edge of Visual Effects. Aurum Press. p. 138. ISBN 978-0823079285. The book, published
  with the full co-operation of the creators of the character (Digital Domain), states the image is a screenshot of a televised commercial for the
  character. *Original artist:* ?

- **File:Nintendo.svg** *Source:* https://upload.wikimedia.org/wikipedia/commons/0/0d/Nintendo.svg *License:* Public domain *Contributors:* Transferred from en.wikipedia to Commons by JamesR. *Original artist:* The original uploader was Trisreed at English Wikipedia

- **File:Noctis_Screenshot.png** *Source:* https://upload.wikimedia.org/wikipedia/commons/f/f9/Noctis_Screenshot.png *License:* CC-BY-SA-3.0 *Contributors:* ? *Original artist:* ?

- **File:Nova_(StarCraft).jpg** *Source:* https://upload.wikimedia.org/wikipedia/en/2/26/Nova_%28StarCraft%29.jpg *License:* ? *Contributors:* Screenshot of trailer, copyright held by Blizzard Entertainment *Original artist:* ?

- **File:Nova_in_combat_(StarCraft).jpg** *Source:* https://upload.wikimedia.org/wikipedia/en/f/fe/Nova_in_combat_%28StarCraft%29.jpg *License:* ? *Contributors:* Subject is copyrighted by Blizzard Entertainment, the image was obtained from GameSpot. *Original artist:* ?

- **File:Orange_lambda.svg** *Source:* https://upload.wikimedia.org/wikipedia/commons/8/8f/Orange_lambda.svg *License:* Public domain *Contributors:* Own work *Original artist:* ineligible

- **File:Portal-puzzle.svg** *Source:* https://upload.wikimedia.org/wikipedia/en/f/fd/Portal-puzzle.svg *License:* Public domain *Contributors:* ? *Original artist:* ?

- **File:Question_book-new.svg** *Source:* https://upload.wikimedia.org/wikipedia/en/9/99/Question_book-new.svg *License:* Cc-by-sa-3.0 *Contributors:* Created from scratch in Adobe Illustrator. Based on Image:Question book.png created by User:Equazcion *Original artist:* Tkgd2007

- **File:Role-playing_video_game_icon.svg** *Source:* https://upload.wikimedia.org/wikipedia/commons/7/7b/Role-playing_video_game_icon.svg *License:* LGPL *Contributors:* Self made from public domain images http://openclipart.org/detail/92041 - http://openclipart.org/detail/91993 & LGPL image File:Gamepad.svg *Original artist:* User:JohnnyMrNinja from images made by yves_guillou & w:David Vignoni

- **File:Sadness_wii_logo.jpg** *Source:* https://upload.wikimedia.org/wikipedia/en/3/31/Sadness_wii_logo.jpg *License:* Fair use *Contributors:* The logo is from the http://www.cubed3.com/news/5043/2/ website. http://www.cubed3.com/news/5043/2/ *Original artist:* ?

- **File:Samsung_ATIV_Q.jpg** *Source:* https://upload.wikimedia.org/wikipedia/commons/5/56/Samsung_ATIV_Q.jpg *License:* CC BY 2.0 *Contributors:* Flickr: ATIV Q_Front_Blue *Original artist:* Samsung Belgium

- **File:Sega-Saturn-Console-Set-Mk1.jpg** *Source:* https://upload.wikimedia.org/wikipedia/commons/a/a8/Sega-Saturn-Console-Set-Mk1.jpg *License:* Public domain *Contributors:* Own work *Original artist:* Evan-Amos

- **File:Shenmueonline_logo.gif** *Source:* https://upload.wikimedia.org/wikipedia/en/a/a1/Shenmueonline_logo.gif *License:* Fair use *Contributors:* ? *Original artist:* ?

- **File:Software_spanner.png** *Source:* https://upload.wikimedia.org/wikipedia/commons/8/82/Software_spanner.png *License:* CC-BY-SA-3.0 *Contributors:* Transferred from en.wikipedia; transfer was stated to be made by User:Rockfang. *Original artist:* Original uploader was CharlesC at en.wikipedia

- **File:Sonic_X-treme_engine_test_screenshot.png** *Source:* https://upload.wikimedia.org/wikipedia/en/9/96/Sonic_X-treme_engine_test_screenshot.png *License:* Fair use *Contributors:* Self-made. *Original artist:* ?

- **File:StarCraft_Ghost_title.jpg** *Source:* https://upload.wikimedia.org/wikipedia/en/d/df/StarCraft_Ghost_title.jpg *License:* Fair use *Contributors:* The logo may be obtained from Blizzard Entertainment. *Original artist:* ?

- **File:Star_empty.svg** *Source:* https://upload.wikimedia.org/wikipedia/commons/4/49/Star_empty.svg *License:* CC BY-SA 2.5 *Contributors:* Made with Inkscape from Stars615.svg: <a href='//commons.wikimedia.org/wiki/File:Stars615.svg' class='image'><img alt='Stars615.svg' src='https://upload.wikimedia.org/wikipedia/commons/thumb/7/72/Stars615.svg/96px-Stars615.svg.png' width='96' height='17' srcset='https://upload.wikimedia.org/wikipedia/commons/thumb/7/72/Stars615.svg/144px-Stars615.svg.png 1.5x, https://upload.wikimedia.org/wikipedia/commons/thumb/7/72/Stars615.svg/192px-Stars615.svg.png 2x' data-file-width='640' data-file-height='110' /></a>. *Original artist:* This vector image was created with Inkscape by Conti from the original images by RedHotHeat, and then manually edited.

- **File:Star_full.svg** *Source:* https://upload.wikimedia.org/wikipedia/commons/5/51/Star_full.svg *License:* Public domain *Contributors:* Made with Inkscape from Image:Stars615.svg. *Original artist:* User:Conti from the original images by User:RedHotHeat

- **File:Star_half.svg** *Source:* https://upload.wikimedia.org/wikipedia/commons/8/81/Star_half.svg *License:* CC BY-SA 2.5 *Contributors:* Made with Inkscape from Image:Stars615.svg. *Original artist:* User:Conti

- **File:Stonekeep_gameplay.png** *Source:* https://upload.wikimedia.org/wikipedia/en/4/47/Stonekeep_gameplay.png *License:* Fair use *Contributors:* Self-made. *Original artist:* Interplay Entertainment

- **File:Sxtreme-jadegully.jpg** *Source:* https://upload.wikimedia.org/wikipedia/en/2/29/Sxtreme-jadegully.jpg *License:* Fair use *Contributors:* Secrets of Sonic Team *Original artist:* ?

- **File:Symbol_book_class2.svg** *Source:* https://upload.wikimedia.org/wikipedia/commons/8/89/Symbol_book_class2.svg *License:* CC BY-SA 2.5 *Contributors:* Mad by Lokal_Profil by combining: *Original artist:* Lokal_Profil

- **File:TF2_BLU_base.jpg** *Source:* https://upload.wikimedia.org/wikipedia/en/0/08/TF2_BLU_base.jpg *License:* ? *Contributors:* <a data-x-rel='nofollow' class='external text' href='http://www.valvesoftware.com/publications/2007/NPAR07_IllustrativeRenderingInTeamFortress2.pdf'>Illustrative Rendering in *Team Fortress 2*</a>. The copyright is held by the Valve Corporation. *Original artist:* ?

- **File:TF2_Group.jpg** *Source:* https://upload.wikimedia.org/wikipedia/en/5/59/TF2_Group.jpg *License:* ? *Contributors:* The image may be obtained from the Valve Corporation *Original artist:* ?

- **File:TF2_RED_base.jpg** *Source:* https://upload.wikimedia.org/wikipedia/en/2/23/TF2_RED_base.jpg *License:* ? *Contributors:* <a data-x-rel='nofollow' class='external text' href='http://www.valvesoftware.com/publications/2007/NPAR07_IllustrativeRenderingInTeamFortress2.pdf'>Illustrative Rendering in *Team Fortress 2*</a>. The copyright is held by the Valve Corporation. *Original artist:* ?

- **File:THEY_from_Games_Convention_2007.jpg** *Source:* https://upload.wikimedia.org/wikipedia/en/1/13/THEY_from_Games_Convention_2007.jpg *License:* Fair use *Contributors:* http://www.they-thegame.com/ promotional screenshot *Original artist:* ?

- **File:THEY_logo.jpg** *Source:* https://upload.wikimedia.org/wikipedia/en/e/e1/THEY_logo.jpg *License:* Fair use *Contributors:* http://www.they-thegame.com/ *Original artist:* ?

- **File:Team_Fortress_2_Screenshot.jpg** *Source:* https://upload.wikimedia.org/wikipedia/en/7/75/Team_Fortress_2_Screenshot.jpg *License:* ? *Contributors:* An in-game screenshot, taken by *PC Gamer UK*. The copyright is held by the Valve Corporation. *Original artist:* ?

- **File:Tf2_oldstyle.jpg** *Source:* https://upload.wikimedia.org/wikipedia/en/3/30/Tf2_oldstyle.jpg *License:* ? *Contributors:* The image may be obtained from the Valve Corporation. *Original artist:* ?

- **File:TheGrinder.jpg** *Source:* https://upload.wikimedia.org/wikipedia/en/0/08/TheGrinder.jpg *License:* Fair use *Contributors:* http://media.wii.ign.com/media/143/14352829/img_6789574.html *Original artist:* ?

- **File:This_Is_Vegas_Logo.jpg** *Source:* https://upload.wikimedia.org/wikipedia/en/2/2f/This_Is_Vegas_Logo.jpg *License:* Fair use *Contributors:* Kotaku write-up *Original artist:* ?

- **File:Tsutenkaku2.jpg** *Source:* https://upload.wikimedia.org/wikipedia/commons/f/f8/Tsutenkaku2.jpg *License:* CC-BY-SA-3.0 *Contributors:* Momopy *Original artist:* Momopy

- **File:Ugo_Volt.jpg** *Source:* https://upload.wikimedia.org/wikipedia/en/e/ea/Ugo_Volt.jpg *License:* Fair use *Contributors:* Ugo Volt, copyright Move Interactive, image from http://www.gamerstek.com/galerias-foto-5293.php *Original artist:* ?

- **File:Valve_logo.svg** *Source:* https://upload.wikimedia.org/wikipedia/commons/a/ab/Valve_logo.svg *License:* Public domain *Contributors:* Transferred from en.wikipedia to Commons. *Original artist:* The original uploader was Pudeo at English Wikipedia

- **File:Warcraft-adventures-boxart.jpg** *Source:* https://upload.wikimedia.org/wikipedia/en/3/33/Warcraft-adventures-boxart.jpg *License:* Fair use *Contributors:* The box/cover art can or could be obtained from Blizzard Entertainment. *Original artist:* ?

- **File:Wiki_letter_w.svg** *Source:* https://upload.wikimedia.org/wikipedia/en/6/6c/Wiki_letter_w.svg *License:* Cc-by-sa-3.0 *Contributors:* ? *Original artist:* ?

- **File:Wiki_letter_w_cropped.svg** *Source:* https://upload.wikimedia.org/wikipedia/commons/1/1c/Wiki_letter_w_cropped.svg *License:* CC-BY-SA-3.0 *Contributors:*

- Wiki_letter_w.svg *Original artist:* Wiki_letter_w.svg: Jarkko Piiroinen

- **File:Wikiquote-logo.svg** *Source:* https://upload.wikimedia.org/wikipedia/commons/f/fa/Wikiquote-logo.svg *License:* Public domain *Contributors:* ? *Original artist:* ?

3.3 Content license